T0188854

EXPERIMENTAL ECOPHYSIOLOGY AND BIOCHEMISTRY OF TREES AND SHRUBS

EXPERIMENTAL ECOPHYSIOLOGY AND BIOCHEMISTRY OF TREES AND SHRUBS

Humberto González Rodríguez, PhD
Ratikanta Maiti, PhD
Ch. Aruna Kumari, PhD

First edition published [2021]

Apple Academic Press Inc.
1265 Goldenrod Circle, NE,
Palm Bay, FL 32905 USA
4164 Lakeshore Road, Burlington,
ON, L7L 1A4 Canada

CRC Press
6000 Broken Sound Parkway NW,
Suite 300, Boca Raton, FL 33487-2742 USA
2 Park Square, Milton Park,
Abingdon, Oxon, OX14 4RN UK

First issued in paperback 2021

© 2021 Apple Academic Press, Inc.

Apple Academic Press exclusively co-publishes with CRC Press, an imprint of Taylor & Francis Group, LLC

Reasonable efforts have been made to publish reliable data and information, but the authors, editors, and publisher cannot assume responsibility for the validity of all materials or the consequences of their use. The authors, editors, and publishers have attempted to trace the copyright holders of all material reproduced in this publication and apologize to copyright holders if permission to publish in this form has not been obtained. If any copyright material has not been acknowledged, please write and let us know so we may rectify in any future reprint.

Except as permitted under U.S. Copyright Law, no part of this book may be reprinted, reproduced, transmitted, or utilized in any form by any electronic, mechanical, or other means, now known or hereafter invented, including photocopying, microfilming, and recording, or in any information storage or retrieval system, without written permission from the publishers.

For permission to photocopy or use material electronically from this work, access www.copyright.com or contact the Copyright Clearance Center, Inc. (CCC), 222 Rosewood Drive, Danvers, MA 01923, 978-750-8400. For works that are not available on CCC please contact mpkbookspermissions@tandf.co.uk

Trademark notice: Product or corporate names may be trademarks or registered trademarks and are used only for identification and explanation without intent to infringe.

Library and Archives Canada Cataloguing in Publication

Title: Experimental ecophysiology and biochemistry of trees and shrubs / Humberto González Rodríguez, PhD, Ratikanta Maiti, PhD, Ch. Aruna Kumari, PhD.
Names: González Rodríguez, Humberto, 1959- author. | Maiti, R. K., 1938- author. | Aruna Kumari, C. H., 1972- author.
Description: Includes bibliographical references and index.
Identifiers: Canadiana (print) 20200202944 | Canadiana (ebook) 20200202995 | ISBN 9781771888370 (hardcover) | ISBN 9780429322266 (ebook)
Subjects: LCSH: Plant ecophysiology. | LCSH: Trees—Physiology. | LCSH: Shrubs—Physiology.
Classification: LCC QK717 .G66 2020 | DDC 581.7—dc23

Library of Congress Cataloging-in-Publication Data

Names: González Rodríguez, Humberto, 1959- author. | Maiti, R. K., 1938- author. | Aruna Kumari, C. H., 1972- author.
Title: Experimental ecophysiology and biochemistry of trees and shrubs / Humberto González Rodríguez, Ratikanta Maiti, Ch. Aruna Kumari.
Description: Palm Bay, Florida : Apple Academic Press, [2021] | Includes bibliographical references and index. | Summary: "The existence and competition of trees and shrubs to sustain and put forth growth under varied environmental conditions is dependent on the interactions that occur between the plant metabolic processes and the prevailing environmental conditions. In order to understand the productivity of trees and shrubs, it is a prerequisite to know the experimental techniques of these vital processes. This volume, Experimental Eco-Physiology and Biochemistry of Trees and Shrubs, provides a comprehensive presentation of this topic. The first part of this book deals with various aspects of experimental ecophysiology and recent research results of studies on plant pigment, epicuticular wax, leaf nutrients, carbon fixation, all supported by literature. The second part of the volume describes various laboratory techniques such as diffusion, imbibition, calorimetry, atomic absorption, mineral nutrition, nutrition analysis of forage, litterfall chemistry, nutrient cycle, etc. The third and fourth parts deal with the advances in the techniques in the development of ecophysiology. The book will serve as an important handbook and resource for students, faculty and teachers, technicians, and researchers involved in forest science dealing with ecophysiology and biochemistry of woody and crop plants"-- Provided by publisher.
Identifiers: LCCN 2020012203 (print) | LCCN 2020012204 (ebook) | ISBN 9781771888370 (hardcover) | ISBN 9780429322266 (ebook)
Subjects: LCSH: Plant ecophysiology. | Trees--Physiology. | Shrubs--Physiology.
Classification: LCC QK717 .G66 2020 (print) | LCC QK717 (ebook) | DDC 581.7--dc23
LC record available at https://lccn.loc.gov/2020012203
LC ebook record available at https://lccn.loc.gov/2020012204

ISBN: 978-1-77188-837-0 (hbk)
ISBN: 978-1-77463-959-7 (pbk)
ISBN: 978-0-42932-226-6 (ebk)

About the Authors

Humberto González Rodríguez, PhD
Faculty Member, Universidad Autonoma de Nuevo Leon,
Facultad de Ciencias Forestales, Mexico

Humberto González Rodríguez, PhD, is a faculty member at the Universidad Autonoma de Nuevo Leon, Facultad de Ciencias Forestales (School of Forest Sciences), Nuevo Leon, Mexico. He received his PhD in Plant Physiology from Texas A&M University under the advice of Dr. Wayne R. Jordan and Dr. Malcolm C. Drew. He is currently working on water relations and plant nutrition in native woody trees and shrubs, northeastern Mexico. In addition, his research includes nutrient deposition via litterfall in different forest ecosystems. Dr. Rodríguez teaches chemistry, plant physiology, and statistics.

Ratikanta Maiti, PhD, DSc
Visiting Research Scientist, Universidad Autonoma de Nuevo Leon,
Facultad de Ciencias Forestales, Mexico

Ratikanta Maiti, PhD, DSc, was a world-renowned botanist and crop physiologist. He worked for nine years on jute and allied fibers at the former Jute Agricultural Research Institute (Indian Council of Agricultural Research), India, and also worked as a plant physiologist on sorghum and pearl millet at the International Crops Research Institute for the Semi-Arid Tropics for 10 years. He also worked for more than 20 years as a professor and research scientist at three different universities in Mexico. In addition, he worked for six years as a Research Advisor at Vibha Seeds, Hyderabad, India, and as a Visiting Research Scientist for five years at the Facultad de Ciencias Forestales, Universidad Autonoma de Nuevo Leon, Mexico. As the author of more than 40 books and about 500 research papers, he won several international awards, including an Ethno-Botanist Award (USA) sponsored by Friends University, Wichita, Kansas, the United Nations Development Programme; a senior research scientist award offered by Consejo Nacional de Ciencia y Tecnología (CONACYT), Mexico; and a gold medal from

India 2008 offered by ABI. He was Chairman of the Ratikanta Maiti Foundation and chief editor of three international journals. Dr. Maiti died in 2019.

Ch. Aruna Kumari, PhD

Assistant Professor, Department of Crop Physiology,
Agricultural College, Jagtial, Professor Jayashankar Telangana State
Agricultural University (PJTSAU), India

Ch. Aruna Kumari, PhD, is an Assistant Professor in the Department of Crop Physiology at Agricultural College, Jagtial, Professor Jayashankar Telangana State Agricultural University (PJTSAU), India. She has six years of teaching experience at PJTSAU and seven years of research experience in varied Indian Council of Agricultural Research institutes as well as at Vibha Seeds. She was a recipient of a Council of Scientific & Industrial Research fellowship during her doctoral studies and was awarded a Young Scientist Award for Best Thesis Presentation at the National Seminar on Plant Physiology. She teaches courses on plant physiology and environmental science and has taught seed physiology and growth, yield, and modeling courses. She has written chapters for several books and is a co-editor of several books. She is also one of editors of the book Glossary in Plant Physiology and an editor of several international books, including *Advances in Bio-Resource and Stress Management; Applied Biology of Woody Plants; An Evocative Memoire: Living with Mexican Culture, Spirituality and Religion; and Gospel of Forests.* She has published over 50 research articles in national and international journals. Her field of specialization is seed dormancy of rice and sunflower.

Contents

Contributors

Sameena Begum, MSc
Researcher, Department of Genetics and Plant Breeding, College of Agriculture,
Rajendranagar, Professor Jayashankar Telangana State Agricultural University (PJTSAU), India

Natalya Sergeevna Ivanova
Ural State Forest Engineering University, Sibirskiytrakt, 37, Yekaterinburg, 620100,
Russia; Botanical Garden of the Ural Branch of the Russian Academy of Sciences,
8th March Str., 202a, Yekaterinburg, 620144, Russia, E-mail: i.n.s@bk.ru

Ch. Aruna Kumari, PhD
Assistant Professor, Department of Crop Physiology, Agricultural College, Jagtial,
Professor Jayashankar Telangana State Agricultural University (PJTSAU), India

Ratikanta Maiti, PhD, DSc
Formerly Botanist and Crop Physiologist; Visiting Research Scientist, Universidad Autonoma de
Nuevo Leon, Facultad de Ciencias Forestales, Nuevo Leon, Mexico

Humberto González Rodríguez, PhD
Faculty Member, Universidad Autonoma de Nuevo Leon, Facultad de Ciencias Forestales
(School of Forest Sciences), Nuevo Leon, Mexico

Ekaterina Sergeevna Zolotova
Researcher, Zavaritsky Institute of Geology and Geochemistry, Ural Branch,
Russian Academy of Science, 15 Akademika Vonsovskogo Street, Yekaterinburg – 620016, Russia,
E-mail: afalinakate@gmail.com

Abbreviations

ABA	abscisic acid
ADF	acid detergent fiber
ADL	acid detergent lignin
ANOVA	analysis of variance
ANOVAR	analysis of variance with repeated measures
ANPP	aboveground net primary productivity
AOAC	Association of Official Analytical Chemists
ASTM	American Society for Testing and Materials
ATP	adenosine triphosphate
AVIRIS	airborne visible infrared imaging spectrometer
BNPP	belowground NPP
C	carbon
CF	charcoal-filtered
CHPM	compensation heat pulse method
CO_2	carbon dioxide
CP	crude protein
DM	dry matter
DRIFT	diffuse reflectance infrared Fourier transform spectroscopy
DW	dry weight
GIS	geographical information system
H^+	hydrogen ion
HFD	heat field deformation
HRM	heat ratio method
IC	ion chromatography
ICP-MS	inductively coupled plasma mass spectrometry
ITV	intraspecific trait variation
LA	leaf area
LL	leaf length
LMA	leaf mass per area
M	molar solution
MANOVAR	multivariate analysis of variance with repeated measures
MPTs	multipurpose trees and shrubs

N	nitrogen
NDF	neutral detergent fiber
NMR	nuclear magnetic resonance
NP split-plot	nonparametric split-plot analysis
O_2	oxygen
OM	organic matter
PC	contents of pipette
PL	pipette of liberation
PLS	partial least squares
PNUE	photosynthetic n-use efficiency
SLA	specific leaf area
SOC	soil organic carbon
SOM	soil organic matter
THB	trunk segment heat balance
TRW	tree ring width
VPD	vapor pressure deficit

Symbols

G_w	old-growth oak exhibited higher stomatal conductance
g_{wv}	stomatal conductance of water vapor
K_l	soil-to-leaf hydraulic conductance
p/v	weight per unit of volume
P_n	net photosynthetic rate
v/v	volume per unit of volume
w/w	weight per unit of weight
δD	isotopic composition
ψ	water potential
Ψ_{leaf}	pre-dawn leaf water potential
ψ_p	turgor pressure
ψ_π^0	zero turgor

Preface

Nature has gifted humanity with a wide variety of trees and shrubs distributed across the globe. These have been serving mankind since the beginning of civilization in the form of supplying timber, honey, fruits, food, and many other necessary needs. They act also as habitat to a large number of animals and birds. The growth and development of trees and shrubs and their productivity, in turn, is dependent on various ecophysiological, physiological, and biochemical processes that are occurring. The existence and competition of trees and shrubs to sustain and put forth growth under varied environmental conditions are dependent on the interactions that occur between the plant metabolic processes and the prevailing environmental conditions. The presence of leaf pigments, leaf epicuticular wax, leaf macro and micronutrients, the carbon sequestration capacity or the carbon fixation ability, the wood chemical compositions, the fiber and lignin contents, the litterfall, etc., play a major role in determining the development and productivity of trees in varied environments. In order to understand the productivity of trees and shrubs, it is a prerequisite to know the experimental techniques of these vital processes, which are rarely available in the literature, though the results are published in journals and books including ours.

The first part of this book deals with various aspects of experimental ecophysiology and recent research results of aspects such as plant pigment, epicuticular wax, leaf nutrients, carbon fixation, all supported by the literature. The second part of this book describes various laboratory techniques for students and researchers such as diffusion, imbibition, calorimetry, atomic absorption, mineral nutrition, nutrition analysis of forage, litterfall chemistry, nutrient cycle, etc. The third and fourth parts deal with the advances in the techniques in the development of ecophysiology.

PART I

EXPERIMENTAL ECOPHYSIOLOGY AND BIOCHEMISTRY

INTRODUCTION

The experimental ecophysiology and biochemistry mention only the up to date research advances of a few aspects of experimental ecophysiology, conclusion, and research needs in Part I, while it discusses the techniques in detail in Part II. Each aspect is stated in the following sequence viz., introduction including literature surveys, methodology, recent results and conclusion, research needs, and references.

The research studies included in this part were undertaken at the experimental station of the Facultad de Ciencias Forestales, Universidad Autonoma de Nuevo Leon, located in the municipality of Linares (24°47′N; 99°32′W), at an elevation of 350 masl. The climate at this locality is mostly subtropical or semiarid. The summers are warm, where the mean monthly air temperature shows a large variation of 14.7°C in January to 23°C in August. Often, the temperatures may go up to 45°C in summer. The average annual precipitation of this area is around 805 mm and it mostly exhibits the bimodal type of rainfall distribution. The Tamaulipan thornscrub or woodlands are major vegetation types in the area (González-Rodríguez et al., 2011). Prevailing soils are deep, dark-gray, lime-gray, lime-clay vertisols, rich in montmorillonite, which shrinks and swells noticeably in response to changes in soil water content (INEGI, 2002). For physiology details please refer 'Plant Physiology' by Taiz and Zeiger (2002).

The techniques used and results obtained are discussed briefly in the following chapters.

CHAPTER 1

Leaf Pigments

1.1 INTRODUCTION

Plants have different types of pigments that perform a significant role in the photochemical reactions, plant metabolism, and in the physiological performance of plants. The chlorophyll pigments (a, b) of chloroplasts play a most important role in the capture of light energy via the photochemical process of photosynthesis leading to the synthesis of the carbohydrate, i.e., converting light energy into the chemical energy (Taiz and Zeiger, 2002).

Apart from chlorophyll pigments, plants also possess pigments like carotenoids, xanthophylls; flavonoids, etc., involved in the protection of chlorophyll pigments from photooxidation, imparts colors to flowers and fruits, act as antioxidants, and take part in the other plant metabolic processes. Each leaf pigment has varied structural characteristics, the contents of which act as indices of protection against the environmental conditions that the plants come across during the course of their growth and development. It is well known that both chlorophyll pigments and carotenoids play a major role in the higher plant photosynthetic processes.

Carotenoids are the naturally occurring fat-soluble pigments found in algae, plants, and also in some of the photosynthetic bacteria, also playing a role in the photosynthetic process. Further, in some non-photosynthetic bacteria, these pigments are found to have a protective role against the damage by light and oxygen (Biswel, 1995; Gitelson et al., 1999). Carotenoids are also known to occur in animals, though animals lack the ability to synthesize them, they obtain and include carotenoids from their diets. These produce bright coloration, serve as antioxidants, and also function as a source of vitamin A activity in animals (Britton, 1995). Besides the above, they also play a role in plant reproduction owing to their capability to attract pollinators because of the presence of bright coloration and also assist in seed dispersal (Yeum and Russell, 2002).

Uvalle Sauceda et al. (2008) observed the presence of variations in chlorophyll a and b and carotenoids according to the seasons. These seasonal variations in these pigment contents were observed in the native shrubs of northeastern Mexico also. These variations in the contents of these pigments also exhibited remarkable variations between the years and species. The research outcomes of Uvalle Sauceda et al. (2008) have revealed the climate effect on the production of the different pigments in plants. Similarly, Castrullo et al. (2000) have studied the variations in the chlorophyll content in some members of the cultivated and wild species of family Famiaceae.

Though the earlier research studies have shown the variations in the pigments content according to the seasonal variations, these variations on other pigments were not studied in our research. Tatini et al. (2005) have studied the roles of specific xanthophylls in light utilization, particularly the role of flavonoids in the integrated mechanism of respiration and photoprotection. Similarly, Kalituho et al. (2007) studied the roles of specific xanthophylls in light utilization, in *Arabidopsis thaliana* at high solar radiation.

The native shrubs and trees existing in the semiarid regions of the northeastern Mexico are used as main resources of food for a broad range of ruminants and white-tailed deer (Ramirez, 1999). Apart from providing food, these are also important in the supply of high-quality fuel and timber for fencing and construction (Reid et al., 1990; Fullbright et al., 1991). Though most of the species in this region are economically important, their growth is also under the influence of the prevailing climatic conditions. The variations in the climatic conditions may bring about the difference in the production of the photosynthetic pigments. Keeping this in view, a study was undertaken in 37 native trees and shrubs of the northeastern Mexico to determine variations in leaf pigments content in summer and in the winter season.

1.2　MATERIALS AND METHODS

1.2.1　PLANT MATERIAL

Thirty-seven native plant species were randomly selected from an area of 2,500 m^2 (50 m × 50 m) representative of the thornscrub. The area was selected from the undisturbed thornscrub plot, which is situated at the research site. Most of the plant species of the thornscrub have several uses. They are used as sources of forage for the domestic livestock and wildlife, fuelwood, charcoal, timber for construction, fences, medicine, agroforestry, and reforestation practices in disturbed sites (Reid et al., 1990).

1.2.2   TISSUE SAMPLING PROCEDURES AND DETERMINATION OF PLANT PIGMENTS

The terminal shoots with fully expanded leaves were sampled from the middle side of five plants (replicates) of each species. The sampled leaves of each species were kept in plastic bags and stored in icebox under dark conditions till the end of the sampling of leaves from all the 37 tree species. These cooled samples were then taken to the laboratory for the pigment contents analysis. The pigment contents were ascertained from these samples within 12 h after collections. For the chlorophyll a, chlorophyll b, and carotenoids content estimation, 5.0 g fresh weight of leaf tissue were extracted in 80% (v/v) aqueous acetone. The extracts were then vacuum filtered through a Whatman No.1 filter paper. Later, the pigment measurements were determined spectrophotometrically in 50 ml of plant pigment sample extract with a Perkin-Elmer UV/VIS Spectrophotometer (Model Lambda 25). Absorbances of chlorophyll a, chlorophyll b, and carotenoids extracts were determined at wavelengths of 663, 645 and 470 nm, respectively. Results are described on a fresh weight basis (mg of plant pigment per g fresh weight). Total chlorophyll (a+b), chlorophyll ratio (a to b) and chlorophyll (a+b) to carotenoids ratio were also determined. The procedure is briefly outlined in the flow diagram in Figure 1.1.

1.3   RESULTS AND CONCLUSIONS

Gonzalez-Rodriguez et al. (2017) studied chlorophyll a, chlorophyll b, and carotenoids contents of 37 species of trees and shrubs in the summer season, in Linares, Northeast of Mexico. Among the species, large variations were observed in the contents of chlorophyll (a, b, and total) and also in carotenoids.

For example, Chlorophyll a was minimum (around 0.6 mg) in *Leucophyllum frutescens, Acacia berlandieri* and maximum (1.8 mg) in *Leucaena leucocephala,* while the Chlorophyll b was minimum in *Forestiera angustifolia, Acacia berlandieri, Acacia palmeri, Leucophyllum frutescens* (0.1 to 0.2 mg). However, in *Leucaena leucocephala* a maximum of 0.4 mg of chlorophyll b was observed. Similarly, the Carotenoids content was also minimum (around 0.2 mg) in *Eysenhardtia polystachya, Guaiacum angustifolium, Leucophyllum frutescens,* and others. Though the total chlorophyll content having minimum values (around 0.5 mg) was found in *Leucophyllum frutescens, Helietta parvifolia, Lantana macropoda, Forestiera angustifolia, Croton suaveolens.* A maximum value (around 2 mg) was found in *Leucaena leucocephala.* It was observed further of chlorophyll (a:b) ratio (around 7) was seen in *Forestiera angustifolia, Parkinsonia aculeata* followed by *Diospyros texana* (around 6). The total chlorophyll/carotenoids ratio was maximum in *Parkinsonia aculeata* (nearing 8), while minimum value (around 1) was found in *Berberis trifoliate.*

[Select fully expanded leaves from middle five plants]

↓

[Store the leaves in icebox under dark conditions till analysis]

↓

[Extract five grams of leaf tissue with 80% (v/v) aqueous acetone]

↓

[Vaccum filter the extracts with Whatman No. 1 Filter paper]

↓

[Measure the absorbance of Chlorophyll a, b, and carotenoids in 50 ml of extract in UV/VIS Spectrophotometer at wavelengths of 663, 645 and 470 nm, respectively]

↓

[Report the results on a fresh weight basis (mg of plant pigment per g fresh weight)]

FIGURE 1.1   Flow diagram for plant pigments estimation.

Gonzalez Rodriguez et al. (2015) conducted a research on nine species of trees and shrubs of the Tamaulipan thornscrub to assess whether the seasons have an effect on the accumulation of pigments. Their studies have revealed the presence of leaf pigments seasonal variations in two seasons (Summer and Winter, 2015). They also reported that among these nine species of trees and shrubs large variations were also seen in the chlorophyll (a and b, total chlorophyll) and carotenoids contents.

1.4 RESEARCH NEEDS

Research needs to be focused to study the role of woody species having high pigment content on the woody tree's physiology and productivity.

KEYWORDS

- carotenoids
- chlorophyll
- flavanoids
- photooxidation
- photoprotection
- plant pigments

REFERENCES

Biswel, B., (1995). Carotenoid metabolism during leaf senescence and its control by light. *J. Photochem. Photobiol. B. Biol., 30*, 3–13.

Britton, G., (1995). Structure and properties of carotenoids in relation to function. *The FASEB J., 9*, 1551–1558.

Castrullo, M. D., Vizcano, E., & Moreno, L. Z., (2000). Chlorophyll content in some cultivated and wild species of Famiaceae. *Biol. Plant., 44*, 423–325.

Fullbright, T. E., Reynold, J. P., Beason, S. L., & Demaaris, S., (1991). Mineral contents of guajillo regrowth following cropping. *J. Range Manage., 44*, 520–522.

Gitelson, A. A., Buschmann, C., Hartman, P., & Lichtenthaler, H. K., (1999). The chlorophyll fluorescence ratio F_{735}/F_{700} as an accurate measure of the chlorophyll content in plants. *Remote Sensing Environ., 69*, 236–302.

González, H., Cantú, S. I., Gómez, M. M. V., & Jordan, W. R., (2000). Seasonal plant water relationships in *Acacia berlandieri. Arid Soil Res. Rehabit., 14*, 343–357.

Gonzalez, R. H., Ratikanta, M., & Aruna, K., (2015). Seasonal influence on pigment production in nine species of trees and shrubs in Linares, northeast of Mexico. *Intern. J. Biores. Stress Manage.*, *6*, 346–351. doi: 10.5958/0976-4038.2015.00059.

González-Rodríguez, H., Cantú-Silva, I., Ramírez-Lozano, R. G., Gómez-Meza, M. V., Sarquis-Ramírez, J., Coria-Gil, N., Cervantes-Montoya, J. R., & Maiti, R. K., (2011). Xylem water potentials of native shrubs from northeastern Mexico. *Acta Agriculturae Scand., Section B - Plant Soil Sci.*, *61*, 214–219.

Gonzalez-Rodriguez, H., Maiti, R., Avendaño, E., Aruna Kumari, Ch., & Sarkar, N. C., (2017). Pigment content (chlorophyll and carotenoids) in 37 species of trees and shrubs in north-eastern Mexico during summer season. *Pak. J. Bot.*, *49*, 173–179.

Kalituho, L., Rech, J., & Jahns, P., (2007). The roles of specific xanthophylls in light utilization. *Planta*, *225*, 423–439.

Kyparissis, A., Drillas, P., & Manetas, Y., 2000. Seasonal fluctuations in photoprotective (xanthophylls cycle) and photoselective (chlorophyll) capacity in eight Mediterranean plant species belonging to two different growth forms. *Aust. J. Plant Physiol.*, *27*, 205–212.

Ramirez, R.G., (1999). Feed resources and feeding techniques of small ruminants under extensive management conditions. *Small. Rumin. Res.*, *34*, 215–230.

Reid, N., Marroquin, J., & Beyer-Minzel, P., (1990). Utilization of shrubs and trees for browsing, fuelwood and timber in Tamaulipan thornscrub in Northeastern Mexico. *For. Ecol. Manage*, *36*, 61–73.

SPP-INEGI, (1986). Síntesis Geográfica del Estado de Nuevo León. Secretaría de Programación y Presupuesto, Instituto Nacional de Geografía e Informática, México.

Taiz, L., & Zeiger, E., (2002). Plant Physiology. 3rd ed. Sunderland. Sinauer Associates. 690 pp.

Tattini, M., Guidi, L, Morassi-Bonzi, L., et al. (2005). On the role of flavonoids in the integrated mechanisms of response of *Ligustrum vulgare* and *Phillyrea latifolia* to high solar radiation. *New Phytol.*, *167*, 457–470.

Uvalle Sauceda, J. I., González-Rodríguez, H., Ramírez Lozano, R. G., Cantú Silva, I., & Gómez Meza, M. V., (2008). Seasonal trends of chlorophylls a and b and carotenoids in native trees and shrubs of Northeastern Mexico. *J. Biol. Sci.*, *8*, 258–267.

Yeum, K., & Russell, R. M., (2002). Carotenoid bioavailability and bioconversion. *Ann. Rev. Nutr.*, *22*, 483–504.

CHAPTER 2

Leaf Epicuticular Wax

2.1 INTRODUCTION

Leaves of many woody trees in Northeastern Mexico contain a waxy coating, commonly denoted as the epicuticular wax. The prevailing environmental conditions may have an influence on the quantity, composition, and morphology of the leaf surfaces waxy coverings (Martin and Juniper, 1970). The waxes are long-chain polymers of lipids. They consist of long-chain paraffins, alcohols, ketones, esters, and free fatty acids in varying proportions. These proportions are under the influence of both the genetic capacity and environmental factors (Silva Fernandes et al., 1964). Several researches have reported that this epicuticular wax increases the reflectance of visible and near-infrared radiation from the leaf surface. This, in turn, reduces net radiation and cuticular transpiration and appears to impart drought resistance of plants (Kurtz, 1950; Ebercon et al., 1977; Hull and Bleckman, 1977). Further, several researchers have reported that these epicuticular waxes on the leaf surfaces of plants also prevent the absorption and penetration of foliar-applied herbicides (Hammerton, 1967; Sharam and Vanden Born, 1970; Wilkinson et al., 1980). In addition, few authors stated that mesquite leaves (*Prosopis* spp.) produce a thick waxy cuticle (Hull, 1970; Meyer et al., 1971). Bleckmann and Hull (1975), stated that in velvet mesquite (*P. velutinu*) wax accumulation increases with maturity, while Meyer et al. (1971) reported that most rapid wax accumulation on honey mesquite (*P. glandulosa*) was seen with early leaf development and expansion.

Bleckmann and Hull (1975) studied the leaf cuticle development of velvet mesquite. They observed the occurrence of definite crystalline wax structures on the youngest leaves also. Jacoby et al. (1990) stated that the wax amount seemed to increase with leaf maturity in *Prosopis* species.

The combined coating of rods and dendritic platelets were present in the epicuticular wax structure of five *Prosopis* species.

Several researches revealed that even under controlled conditions various factors such as light intensity (Juniper, 1960; Reid and Turkey, 1982), photoperiod (Wilkinson, 1972), temperature (Hull, 1958; Reed and Tukey, 1982) and water stress (Skoss, 1955; Bengston et al., 1978; Baker and Procopiou, 1980) had an effect on the epicuticular wax development. Further, Rao and Reddy (1980) have reported that in a semiarid environment the seasonal variations were exhibited in the composition and quantity of shrubs epicuticular waxes. The content and composition varied with the variants of temperature and rainfall. It was also reported that both cuticular and total transpiration seemed to be influenced by wax composition changes.

Most of the research review has well demonstrated the epicuticular wax significance in different physiological functions in plants, for instance, the solar radiation reflectance, imparting of drought resistance, etc. Keeping in view the above, a study was carried out to determine the epicuticular wax variability among woody plant species in the Tamaulipan thornscrub and ultimately to select the species with high epicuticular wax. This study was carried out using 35 woody trees species in the summer season (June 2015) at the experimental station of Facultad de Ciencias Forestales, Universidad Autonoma de Nuevo Leon, located in the municipality of Linares. Fresh leaves were collected from 35 woody tree species of the Tamaulipan thornscrub and their leaflets were separated individually so as to obtain a subsample with an estimated surface area of 100 cm^2 which is determined by a leaf area (LA) meter. The sub-samples thus obtained were gently rinsed in distilled water to get rid of foreign material adhered to the leaf surfaces. These were then air-dried and then placed in a beaker with 40 ml of analytical grade chloroform (99% pure) and preheated to 45°C. After 30s, the chloroform was poured into preweighed foil pans. These were then placed in a well-ventilated fume hood and evaporated to dryness for 24 hours. Foil pans were then reweighed to measure the residual wax amount. The amount of wax for a field sample was the mean of five replications. The wax was calculated and reported on a weight per area basis (g m^{-2}) derived by dividing wax weight by the actual area of the sub-sample unit (Figure 2.1).

[Collect the fresh leaves and obtain a subsample of 100 cm^2
surface area by leaf area meter]

↓

[Rinse the subsample with distilled water to remove dust and
foreign particles and air dry the sample]

↓

[Place the air-dried sample in a beaker containing 40 ml of
analytical grade chloroform (99% pure) and preheated to 45°C]

↓

[After 30 s pour the chloroform into preweight foil pans]

↓

[Place them in a well-ventilated fume hood and evaporate to
dryness for 24 hrs]

↓

[Reweigh the foil pans to quantify the amount of residual wax]

↓

[Calculate the wax content and report it on a weight per area basis (g m^{-2})
derived by dividing wax weight by the actual area of the sub-sample unit]

FIGURE 2.1 Flow diagram for leaf epicuticular wax content estimation.

2.2 RESULTS AND CONCLUSIONS

The 35 species analyzed for epicuticular wax content had shown significant variations in epicuticular wax contents. Maximum tree species exhibited a large variation in the wax contents where the wax contents varied from 11.18 to 702.04 µg/cm^{-2}. The species were selected and classified into high, medium, and low wax content species based on their total epicuticular wax content.

Among these 35 woody tree species some of the species viz., *Forestiera angustifolia* (702.04 µg/cm^2), *Diospyros texana* (607.65 µg/cm^2), *Bernardia myricifolia* (437.53 µg/cm^2), *Leucophyllum frutescens* (388.50 µg/cm^2), *Acacia farnesiana* (373.49 µg/cm^2), and *Cercidium macrum* (308.63 µg/cm^2) have shown the presence of high epicuticular wax contents. Certain species viz., *Lantana macropoda* (294.86 µg/cm^2), *Quercus polymorpha* (199.40 µg/cm^2), *Parkinsonia aculeata* (196.20 µg/cm^2), *Acacia shaffneri* (170.04 µg/cm^2), *Diospyros palmeri* (163.25 µg/cm^2), *Helietta parvifolia* (151.19 µg/cm^2), *Eysenhardtia polystachya* (138.49 µg/cm^2), and *Bumelia celastrina* (132.38 µg/cm^2) exhibited a medium epicuticular wax load on the leaves. Apart from these, the species *Ehretia anacua* (17.58 µg/cm^2), *Karwinskia humboldtiana* (15.47 µg/cm^2), and *Amyris texana* (11.18 µg/cm^2) had a low epicuticular wax content.

In conclusion, it may be stated that from many researches it is clear that the presence of epicuticular wax on the leaf surface has a significant role in the physiological functions and adaptation of trees in an ecosystem. The presence of epicuticular wax on the leaf surfaces enables in a decrease of the transpiration rates because of the solar radiation reflection and also has an influence on other processes as gas exchange, water stress, herbicide resistance, etc. The research findings of the present study have indicated the existence of large variations in the contents of epicuticular wax among these 35 woody tree species, thereby giving us an opportunity to select species for future research with respect to the adaptation of these species to the semiarid environment.

2.3   RESEARCH NEEDS

The species with high contents of epicuticular wax were selected viz., *Forestiera angustifolia* (702.04 µg/cm^2), *Diospyros texana* (607.65 µg/cm^2), *Bernardia myricifolia* (437.53 µg/cm^2), and *Leucophylum frutescens* (388.50 µg/cm^2). It is estimated that these species may be well adjusted to the semi-arid conditions. Forthcoming research needs to be focused on these selected species with special reference to their physiological function and adaptation to water stress. Besides further study is needed on the seasonal epicuticular wax variation of the selected species.

KEYWORDS

- *Bernardia myricifolia*
- *Diospyros texana*
- epicuticular waxes
- herbicide resistance
- paraffin
- *Parkinsonia aculeata*

REFERENCES

Baker, E. A., & Procopiou, J., (1980). Effect of soil moisture status on leaf surface wax yield of some drought-resistant species. *J. Hort. Sci., 55,* 85–87.

Bengtson, C., Larsson, S., & Liljenberg, C., (1978). Effects of water stress on cuticular transpiration rate and amount and composition of epicuticular wax in seedlings of six oat varieties. *Physiol. Plantarum, 44,* 319–324.

Bleckmann, C. A., & Hull, H. M., (1975). Leaf and cotyledon surface ultrastructure of five *Prosopis* species. *J. Arizona Acad. Sci., 10,* 98–105.

Ebercon, A., Blum, A., & Jordan, W. R., (1977). A rapid calorimetric method for epicuticular wax content of sorghum leaves. *Crop Sci., 17,* 179–180.

Hammerton, J. L., (1967). Environmental factors and susceptibility to herbicides. *Weeds, 15,* 330–336.

Hull, H. M., (1956). The effect of day and night temperature on growth, foliar wax content, and cuticle development of velvet mesquite. *Weeds, 6,* 133–142.

Hull, H. M., (1970). Leaf structure as related to absorption of pesticides and other compounds. *Residue Rev., 31,* 1–150.

Hull, H. M., Morton, H. L., & Wharrie, J. R., (1975). Environmental influences on cuticle development and resultant foliar penetration. *Bot. Rev., 41,* 421–452.

Hull, H. M., & Bleckmann, C. A., (1977). An unusual epicuticular wax ultrastructure on leaves of *Prosopis tamarugo* (leguminosae). *Amer. J. Bot., 4,* 1083–1091.

Hull, H. M., Went, F. W., & Bleckman, C. A., (1979). Environmental modification of epicuticular wax structure of Prosopis leaves. *J. Arizona-Nevada Acad. Sci., 14,* 39–42.

Jacoby, P. W., Anslet, R. J., Meadors, C. H., & Huffman, A. X., (1990). Epicuticular wax in honey mesquite: Seasonal accumulation and intraspecific variation. *J. Range Manage., 43,* 347–350.

Juniper, B. E., (1960). Growth, development, and effect of the environment on the ultrastructure of plant surfaces. *J. Linnean. Soc. Bot., 56,* 413–418.

Kurtz, E. B., (1950). The relation of the characteristics and yield of wax to plant age. *Plant Physiol., 25,* 269–278.

Martin, J. T., & Junier, B. E., (1970). *The Cuticles of Plants.* St. Martin's, New York.

Meyer, R. E., Morton, H. L., Haas, R. H., Robison, E. D., & Riley, T. E., (1971). Morphology and anatomy of honey mesquite. *U.S. Dep. Agr. Tech. Bull., 1423,* 186.

Rao, J. V. S., & Reddy, K. R., (1980). Seasonal variation in leaf epicuticular wax of some semiarid shrubs. *Indian J. Exp. Biol.*, *18*, 495–499.

Reed, D. W., & Tukey, H. B., (1982). Light intensity and temperature effects on epicuticular wax morphology and internal cuticle ultrastructure of carnations and Brussels sprouts leaf cuticles. *J. Amer. Soc. Hort. Sci.*, *107*, 417–420.

Sharma, M. P., & Vanden Born, W. H., (1970). Foliar penetration of picloram and 2,4-D in aspen and balsam poplar. *Weed Sci.*, *18*, 57–63.

Silva Fernandes, A. M., Baker, E. A., & Martin, J. T., (1964). Studies on plant cuticle-VI. The isolation and fractionation of cuticular waxes. *Ann. Appl. Biol.*, *53*, 43–58.

Skoss, J. D., (1955). Structure and composition of plant cuticle in relation to environmental factors and permeability. *Bot. Gaz.*, *117*, 55–72.

Wilkinson, R. E., (1972). Sicklepod hydrocarbon response to photoperiod. *Phytochemistry*, *11*, 1273–1280.

Wilkinson, R. E., (1980). Ecotypic variation of *Tamarix pentandra* epicuticular wax and possible relationship with herbicide sensitivity. *Weed Sci.*, *28*, 110–113.

CHAPTER 3

Leaf Nutrients

3.1 INTRODUCTION

The plant growth and productivity of woody trees and shrubs are accredited to the presence of leaves with varying photosynthetic capacities and nutrient contents. Among these species of woody trees and shrubs, an enormous diversity is seen in the growth form, leaf size, and leaf shape and canopy structure. Besides, there also exist some relationships across a vast range of diversity in leaf traits, in defining the carbon fixation strategy among species. The outer canopy leaves, its specific leaf area (SLA, leaf area (LA) per unit mass) appear to be associated with leaf nitrogen per unit dry mass, photosynthesis, and dark respiration sites (Wright et al., 2001).

Plant roots absorb many macro and micronutrients from soil profiles, needed for plant growth and development. Leaves contain these macro and micronutrients in varying amounts. These nutrients also act as a nutrients source for grazing animals in the forest ecosystem. The presence of large variations in leaf traits among species favors nutrient conservation. The nutrients conserved in the leaves permit the short-term rapid growth of these tree or shrub species. Further, it is apparent from several research findings that the species having high nutrient conservation abilities have a long life span, high leaf mass per area (LMA), low nutrient concentrations and low photosynthetic capacity. Therefore, the availability of the nutrients in leaves is essential for effective physiological and biochemical plant functions.

Chapin (1982) made a review on the crop responses nature to nutrient stress and compared these responses to those species which have evolved under more natural conditions. He concentrated his nutritional research studies on nitrogen and phosphorus, because these were the two major macronutrients commonly exhibiting their effect on the plant growth and development. The nutrients available in leaves contribute to a large number of activities involved in metabolism and plant growth. Ample research inputs were undertaken globally on nutrient contents in leaves and their affect on the plant metabolic processes. The nutrients that are accumulated in the leaves of trees or shrubs are closely correlated with the soil habitats' nutrient content.

According to Chapin et al. (1990), nutrient-deficient habitats tend to be dominated by species which are nutrient conserving, while fertile habitats tend to be dominated by those species which have a higher short-term productivity per leaf mass. The nutrient resorption commences as the leaves start aging with time and these nutrients are removed from the aged and senesced leaves before abscission and leaf fall. Thus, these nutrients are again reutilized for facilitating the metabolic processes involved in the growth and development of the developing tissues (leaves, fruits, seeds). Though the process of resorption of nutrients from leaves occurs throughout the life period of a leaf especially when the leaves are fallen (Ackerly and Bazzaz, 1995; Hikosaka et al., 1994), a major phase of resorption occurs just before leaf abscission. This is a highly organized process occurring in most species during leaf senescence (Noodén, 1988). Some of the research findings have indicated that around 50% of the N and P nutrient contents are recycled via resorption (Aerts, 1996). Negi and Singh (1993) have emphasized the existence of active nutrient sinks in the plants exhibiting their control over resorption.

Regel and Marschner (2005) have undertaken a study on nutrient availability and management of nutrients in the rhizosphere of the plant species exhibiting genotypic differences. It is found that a range of mechanisms gets activated on subjecting the plants to nutrient deficiency leading to increased nutrient availability in the rhizosphere when compared with bulk soil. The plants may alter their root morphology, increase nutrient transporters affinity in the plasma membrane, or also exude some of the organic compounds like carboxylates, phenolics, carbohydrates, enzymes, etc., to increase their ability of nutrient absorption from the soil rhizosphere. The chemical changes that occur in the rhizosphere also seem to add to changed abundance and composition of microbial communities. The study revealed that the nutrient efficient genotypes are more adapted to the environments with low nutrients. So, it is evident that the clear understanding of the role of plant-microbe-soil interactions governing the nutrient availability will bring about the enhancement of environmental sustainability.

A strategy for changes in leaf physiology, structure, and nutrient content between species of high- and low-rainfall and high- and low-nutrient habitats was formulated by Wright et al. (2001). In his research findings, he mentioned that in most plants the nutrients are withdrawn by the plants as the leaves advance in age. However, it was observed that the proportions of nutrients resorbed and the residual nutrient concentrations in senesced leaves were quite varying. Thus, a major spectrum of strategic variations in the nutrient resorption was seen in those plant species with a long life span, high leaf mass per unit area, low leaf nutrient concentrations, and low photosynthetic capacity. Further, the estimations of green-leaf and senesced-leaf N and P concentrations have indicated that the leaf nutrient concentrations in green and senesced leaves were positively correlated with leaf length (LL) across all species that were analyzed. In most sites without nitrogen-fixing species, it was observed that the proportional resorption did not explain the difference with soil nutrients. The above research findings have supported the argument that nutrient losses have an effect on the residual nutrient concentration instead of proportional resorption per se.

Lukhele and van Ryssen (2003) have directed their research towards the value of the leaf nutrients of plants used as forage in meeting the nutrient requirements of ruminants and wild animals. The research study carried out on the chemical composition and potential value of a subtropical tree species of *Combretum* in Southern Africa for ruminants has shown that the species was not an appropriate N resource to supplement protein deficiencies in low-quality herbage.

3.2 METHODOLOGY

3.2.1 CHEMICAL ANALYSIS

Mature leaf samples (1.0 g dry weight (DW)) were taken from each plant and shrub species of the Tamaulipan thornscrub selected for estimating the contents of minerals (Cu, Fe, Zn, Ca, Mg, K, and P). In the dried leaf sample, mineral content was estimated by incinerating samples in a muffle furnace at 550°C for 5 hours. The leaf ashes thus obtained were digested in a solution containing HCl and HNO_3, using the wet digestion technique (Cherney, 2000). The nutrient concentrations of Ca (nitrous oxide/acetylene flame), Cu, Fe, Zn, K, and Mg (air/acetylene flame) were estimated by atomic absorption spectrophotometry (Varian, model SpectrAA-200), whereas P was quantified spectrophotometrically with a Perkin-Elmer spectrophotometer (Model Lambda 1A) at 880 nm (Association of Official Analytical Chemists (AOAC), 1997) (Figure 3.1).

[Obtain one gram of leaf sample dry weight]

↓

[Determine the mineral content in the dried leaf sample
by incinerating samples in a muffle furnace at 550°C for 5 hours]

↓

[Digest the leaf ashed samples in a solution containing
HCl and HNO_3, using the wet digestion technique]

↓

[Estimate the nutrient concentrations of Ca
(nitrous oxide/acetylene flame), Cu, Fe, Zn, K, and Mg
(air/acetylene flame) by atomic absorption spectrophotometry]

↓

[Quantify P at 880 nm in a spectrophotometer]

FIGURE 3.1   Flow diagram for nutrient contents determination.

3.3   RESULTS AND CONCLUSIONS

Gonzalez Rodriguez et al. (2017) have undertaken a research study to evaluate six nutrients in the leaves, three macronutrients (K, Mg, P) and three micronutrients (Cu, Fe, Zn) of 37 woody species in Linares, Northeastern of Mexico. The research outcomes have shown the existence of wide variability in the above mineral nutrient contents in the species studied. From the perspective of the results obtained they stated that the species showed large variation in the contents of three macro (K, Mg, P) and micronutrients (Cu, Fe, Zn), thus giving the researchers a chance to select species for high macro and micronutrients. The species having highest nutrient contents such as *Croton suaveolens* with 75.62 mg/g of K and 2.43 mg/g of P, *Lantana macropoda* with 3,71 mg/g of Mg, *Cordia boissieri* with 30.71 µg/g dw of Cu and 280.55 µg/g dw of Fe and *Salix lasiolepis* with 148.86 µg/g dw of Zn content could serve as excellent sources for macro and micronutrient for the ruminants and could be adapted and grown over larger areas.

3.4 RESEARCH NEEDS

It is essential to study the growth, development, and adaptation of the species having the highest nutrient contents to the semi-arid environment.

KEYWORDS

- atomic absorption
- phenolics
- rhizosphere
- spectrophotometry
- Tamaulipan thornscrub

REFERENCES

Ackerly, D. D., & Bazzaz, F. A., (1995). Leaf dynamics, self-shading and carbon gain in seedlings of a tropical pioneer tree. *Oecologia, 101,* 289–298.

Aerts, R., (1996). Nutrient resorption from senescing leaves of perennials: Are their general principles. *J. Ecol., 84,* 597–608.

Association of Official Analytical Chemists (AOAC), (1997). *Official Methods of Analysis.* Washington, DC: AOAC.

Chapin, F. S., (1982). The mineral nutrition of wild plants. *Ann. Rev. Ecol. Syst., 11,* 233–260.

Chapin, F. S., Schulze, H A Mooney, E. D., & Mooney, H. A., (1990). The ecology and economics of storage in plants. *Ann. Rev. Ecol. Syst., 21,* 423–447.

Cherney, D. J. R., (2000). Characterization of forages by chemical analysis: Ch. 14. In: Givens, D. I., Owen, E., Axford, R. F. E., & Ohmed, H. M., (eds.), *Forage Evaluation in Ruminant Nutrition* (pp. 281–300). CABI Publishing, Wallingford, UK.

Gonzalez Rodriguez, H., Maiti, R. K., Kumari, C. A., & Sarkar, N. C., (2017). Woody plant species with high nutritional value, northeastern Mexico. *Intern. J. Biores. Stress Mange., 8,* 450–456.

Hikosaka, K., Terashima, I., & Katoh, S., (1994). Effects of leaf age, nitrogen nutrition and photon flux density on the organization of photosynthetic apparatus in the leaves of a vine (*Ipomoea tricolor* Cav.) grown horizontally to avoid natural shading of leaves. *Planta, Oecologia, 97,* 451–457.

Lukhele, M. S., & van Ryssen, J. B. J., (2003). Chemical composition and potential nutritive value of subtropical tree species in South Africa for ruminants. *S. Afr. J. Anim. Sci., 33,* 132–141.

Negi, G. C. S., & Singh, S. P., (1993). Leaf nitrogen dynamics with particular reference to retranslocation in evergreen and deciduous tree species in Kumaun Himalaya. *Can. J For. Res., 23,* 349–357.

Noodén, L. D., (1988). The phenomena of senescence and aging. In: Noodén, L. D., &. Leopold, A. C., (eds.), *Senescence and Aging in Plants*. Academic Press. doi: https://doi. org/10.1016/B978-0-12-520920-5.X5001-9

Regel, Z., & Marschner, P., (2005). Nutrient availability and management in the rhizosphere: Exploiting genotypic differences. *New Phytol., 168,* 305–312. doi: 10.1111/ j,1469-8137,2005.015558x.

Wright, I. J., Reich, P. B., & Westoby, M., (2001). Strategy shifts in leaf physiology, structure, and nutrient content between species of high- and low-rainfall and high- and low-nutrient habitats. *Funct. Ecol., 15,* 423–434.

CHAPTER 4

Carbon Fixation (Carbon Sequestration)

4.1 INTRODUCTION

Trees and shrubs have the capacity to capture carbon dioxide through the process of photosynthesis through carbon fixation (carbon sequestration) from the atmosphere. Owing to the continuous emission of these toxic gases by the burning of fossil fuels and other human activities, there is a huge increase in carbon dioxide concentrations. Adequate research activities are directed on this aspect.

Alig et al. (2002) through their research findings have stated that the causes associated with increased global warming are the incessant logging of trees, illegal timber harvest and other anthropogenic activities, increased agricultural lands that aroused with the conversion of vast tracts of forest areas. Apart from these, several other activities either directly or indirectly are responsible for the increased carbon dioxide accumulation and other greenhouse gases in the atmosphere. These entire greenhouse gases have also increased tremendously in recent years aggravating the impact of global warming all over the earth's surface in the form of climate change. Global warming is being also increased with the increase in the pollution rates, by the continuous emissions of carbon dioxide from fossil fuels combustion from the factories and burning of woods. The gas that is being accumulated in the atmosphere is of a great threat in causing an increase in pollution in the atmosphere thereby, endangering the security and sustainability of mankind and animals. Several researches have shown that global warming is probable to exhibit its direct effect on climate changes, thereby, reducing crop productivity and enhancing poverty. The countries all over the globe have formulated several strategies for the mitigation of this climate change arising due to increased levels of carbon dioxide and other greenhouse gases. Several technologies are being adopted for the capture of this noxious gas in the form of carbon dioxide capture and carbon sequestration. Though these technologies are easily adopted in the developed countries, the developing countries are lagging behind to adopt these technologies and are unable to afford to bear these high-cost technologies used for carbon dioxide capture. In this aspect, an alternative strategy to mitigate climate change and global warming is to utilize the efficiency of trees and their enormous capacity to capture the stacked carbon dioxide from the atmosphere by the process of photosynthesis. Trees synthesize carbohydrates from this carbon dioxide through photosynthesis and store it as carbon in their biomass.

During the process of photosynthesis, plants take in CO_2 and release the oxygen (O_2) to the atmosphere. This oxygen is made available for our respiration. The plants store carbon and utilize the stored carbon for growth to guide all metabolic functions (Taiz and Zeiger, 1998).

We narrate here a brief review of research undertaken in carbon fixation and its influence on climate change.

Carbon sequestration involves the capturing carbon dioxide (CO_2) from the atmosphere or capturing anthropogenic (human) CO_2 from large-scale stationary sources like power plants prior to its release to the atmosphere. There are two major types of CO_2 sequestration: terrestrial and geologic.

Terrestrial sequestration of carbon dioxide involves all those processes of land management practices leading to the enhancement of the amount of carbon stored both in the soil and plant material for a long period of time.

Geologic sequestration of carbon dioxide is another process of carbon sequestration process (CCS) process. Unlike terrestrial, or biologic, in geologic sequestration, the carbon is stored through agricultural and forestry practices. In Geologic, sequestration the carbon dioxide is injected deep underground. In the underground, the carbon dioxide is retained permanently. In order to meet the carbon dioxide standards released into the atmosphere, new power plants can employ carbon capture and sequestration technologies.

Carbon fixation in trees involves a micro-optimization process. This process leads to the accumulation of carbon in plant organs and stored as its biomass. Two alternative economic analog models of carbon fixation in trees were demonstrated by Hof et al. (1990). In these ecological studies in the first model, he explained that the plant functions as a maximizer of net carbon gain (a profit analog); whereas in his second model he considers that plant functions as carbon 'revenue.' Carbon gain and carbon revenue are the minimum of two functions in his ecological model. This includes the carbon gain to leaf and root biomass, respectively. In both of these models, the limiting factors are the leaves and roots. Similarly, Coleman et al. (1995) have undertaken a research on photosynthetic productivity of aspen clones. The research findings revealed the variability in aspen clones sensitivity to troposphere ozone. He observed through his research that environments have an effect on the aspen clone's activity.

It is clearly evident through several researches that life on this earth is sustaining mainly because of the carbon dioxide that is being fixed into the living matter of the biosphere with geochemistry.

The details of the early evolutionary history of biological carbon-fixation and its emergence were explained through models designed by Braakman and Smith (2012). They explained that biological carbon fixation has contributed to all the modern pathways in a single ancestral form itself. Their results revealed the innovative processes of carbon fixation which led to the foundations of research for the existence of the present-day major early divergences in the life of trees. Thus, their research findings have given clarification on the integration of the metabolic and phylogenetic constraints of the biological carbon fixation. Likewise in the research studies undertaken Okimoto et al. (2013) on net carbon fixation of a mangrove tree *Rhizophora apiculate*, representative of South East Asia has shown that its net carbon fixate was 2.5–30.5 Mg C ha^{-1} yr^{-1}. The estimated values were found to be significantly higher to the results which have been acquired through the growth curve analysis method.

Several researches have indicated the climatic changes that are expected to be seen in the future, due to the release of carbon dioxide and other greenhouse gases from the factories, coal mine areas and other sources. To reduce the climate change effect a strategic technology accessible to the mankind is to utilize the inherent capacity of trees to capture this carbon load of the atmosphere into their biomass. In this direction, the EPA has adopted several strategies to alleviate the effects of global warming and one such novel technique is the reforestation in and around the factories, mining areas, etc. Reforestation is taken on a massive scale in order to reduce the increased load of carbon in the atmosphere.

Though several researches have shown that carbon dioxide sequestration is one of the efficient strategies available to manage the future climate change, Keller et al. (2003) have developed analytical models to explore the problem in a simple and a realistic manner. To understand the carbon sequestration and its effectiveness to manage the future climatic changes in more optimal economic growth they have developed a simple analytical model and a numerical optimization model. Through their research find-ings, they have determined that carbon sequestration is not the only perfect substitute for mitigating future climatic changes as there is also the possi-bility of leaks of carbon dioxide back to the atmosphere during the carbon dioxide production process and would thus impose more to the future costs of mitigation strategies likely to be adopted.

The carbon concentration in different organs of the *Larix olgensis* tree with varying ages, between 7.5 and 46 years were evaluated by dry combustion method with a Vario EL III element analyzer by Fu et al. (2013) in North-Eastern China. The research findings showed the exis-tence of the variability in the weighted mean carbon concentration by biomass of aboveground tree organs and belowground organs. The mean carbon concentration in the aboveground parts was found to be around 48.15%. Further, it was noticed that in the organs of aboveground parts of *Larix olgensis* the carbon concentration exhibited a decreasing trend from a living branch, bark, foliage to dead branch and stem. However, such significant variations in the carbon concentration were not observed in the trees of different ages that were analyzed.

The carbon content in live wood is one of the criteria adopted to quan-tify the C stocks of the tropical forest trees. In a 50 ha forest plot on Barro Colorado Island, Panama, Martin & Thomas (2011) studied the carbon contents and its variability in the live wood of the tropical tree species. They observed that the wood C content in these species varied from 41.9 to 51% and assumed that the generic C fractions in the tropical wood have overestimated the forest C stocks by ~3.3 to 5.3% higher. This has led to an overestimate of 4.1–6.8 Mg C ha^{-1} in the above plot area.

4.2 MATERIALS AND METHODS

This study was conducted at the experimental station of Facultad de Ciencias Forestales, Universidad Autonoma de Nuevo Leon, located in the municipality of Linares (24°47′N; 99°32′W), at an elevation of 350 m.

4.2.1 CHEMICAL ANALYSIS

The leaves of 30 woody and shrubs were collected during autumn and placed to dry in the newspaper for a week. The leaves were separated from the rest of the plant and were passed twice through a mesh of 1 mm × 1 mm using a Thomas Wiley mill and then dried for more than three days at 65°C in an oven (Precision model 16EG) to remove moisture from the sample and later these were placed in desiccators. A 2.0 mg of the sample was weighed in an AD 6000 Perkin-Elmer balance in a vial of tin, bent perfectly. This was placed in a CHONS analyzer Perkin Elmer Model 2400 for determining carbon, hydrogen, and nitrogen. Crude protein content was determined by multiplying nitrogen content by 6.25 (Figure 4.1).

4.3 RESULTS AND CONCLUSIONS

In the context of the above literature survey, a study was undertaken to estimate carbon fixation (sequestration) of 30 woody species mostly trees in Linares, northeast of Mexico with the main objective to select species with carbon fixation for a recommendation of plantation in carbon dioxide polluted areas.

Maiti et al. (2015) undertook a study to determine carbon fixation (carbon sequestration), nitrogen, C/N and protein contents of 30 woody trees and shrubs in Linares, Northeast Mexico with a view to select species with high carbon fixation (carbon content) and nitrogen content. Only some species with high carbon fixation selected in this study were *Leucophyllum frutescens* 49.97% *Forestiera angustifolia* 49.47%, *Blumelia celastrina* 49.25%, and *Acacia berlandieri* 49.18%, *Sideroxylon celastrina* 49.25%. Some of the species selected with moderately high carbon fixation ability were *Acacia rigidula* 48.23%, *Acacia farnesiana* 46.17%, *Gymnospermum glutinosaum* 46.13%, *Croton suaveolens* 45.17%, and *Sargentia gregii* 44.07%. Through our research findings, it was observed that the trees exhibit a large variability in the carbon fixation capacities, a few of these species with high carbon fixation rates could be suggested for plantation in carbon dioxide polluted areas so as to enable in reduction of the carbon load from the atmosphere at least to some extent. Further, these trees could also serve as good sources of energy by the supply of wood charcoal to the forest dwellers. The tree species not only exhibited variability in the carbon fixation, variability was also observed with the nitrogen and crude protein contents. The nitrogen content in these ranged from 2 to 5%, while the crude protein content is around 11 to 37%. The C/N content ranged from 7 to 36%, representing the utility of these species to serve as a nutritional source of forage to the animals.

[Procure the leaves and dry them for a week]

↓

[Pass the leaves twice through a mesh of 1 × 1 mm
using a Thomas Wiley mill]

↓

[Dry the leaves for more than three days at 65°C in an oven]

↓

[Weigh 2.0 mg of the sample in an AD 6000 Perkin-Elmer balance
in a vial of tin, bent perfectly]

↓

[Place the sample in a CHN analyzer
for determining carbon, hydrogen and nitrogen]

↓

[Calculate the carbon and nitrogen contents (% dry mass basis) in 0.020 g of
milled and dried leaf tissue by using a CHN analyzer]

↓

[Determine crude protein content by multiplying the nitrogen content by 6.25]

FIGURE 4.1   Flow diagram for determining carbon, nitrogen and crude protein contents.

KEYWORDS

- *Blumelia celastrina*
- **carbon dioxide**
- *Gymnospermum glutinosaum*
- *Larix olgensis*
- *Leucophyllum frutescens*

REFERENCES

Alig, R. J., Adams, D. M., & McCarl, B. A., (2002). Projecting impacts on global climate change on the US forest and agriculture sectors and carbon budgets. *Forest Ecol. Manage., 169,* 3–14.

Braakman, R., & Smith, E., (2012). The emergence and early evolution of biological carbon-fixation. *PLoS Comput. Biol., 8,* e1002455.

Cherney, D. J. R., (2000). Characterization of forages by chemical analysis: Ch. 14. In: Givens, D. I., Owen, E., Axford, R. F. E., & Ohmed, H. M., (eds.), Forage Evaluation in Ruminant Nutrition (pp. 281–300). CABI Publishing, Wallingford, UK.

Coleman, M. D., Isebrands, J. G., Dickson, R. E., & Karnosky, D. F., (1995). Photosynthetic productivity of aspen clones varying in sensitivity to tropospheric ozone. Tree Physiol., 15, 585–592.

Fu, Y. Y., Wang, X., & Sun, Y., (2013). Carbon concentration variability of Larix olgensis in North-Eastern China. *Advance J. Food Sci. Technol., 5,* 627–632.

Hof, J., Rideout, D., & Binkley, D., (1990). Carbon fixation in trees as a micro-optimization process: An example of combining ecology and economics. *Ecological Economics, 2,* 243–256.

Keller, K., McInerney, D., & Bradford, D. F., (2008). Carbon dioxide sequestration: how much and when?. *Climatic Change, 88,* 267.

Maiti, R., Rodriguez, H. G., & Kumari, Ch. A., (2015). Trees and shrubs with high carbon fixation/concentration. *Forest Res. Open Access,* S1, 003. doi: 10.4172/2168-9776. S1-003

Martin, A. R., & Thomas, S. C., (2011). A reassessment of carbon content in tropical trees. *Plos One, 6,* e23533.

Okimoto, Y., Nose, A., Oshima, K., Tateda, Y., & Ishii, T., (2013). A case study for an estimation of carbon fixation capacity in the mangrove plantation of *Rhizophora apiculate* trees in Trat, Thailand. *Forest Ecol. Manage., 310,* 1016–1026.

Taiz, L., & Zeiger, E., (1998). Plant Physiology. Sinauer Associates, Inc. Publishers. Sunderland, Massachussets, USA. 792 p.

CHAPTER 5

Wood Carbon and Nitrogen

5.1 INTRODUCTION

Timber yielding plants yield wood as one of their vital economical product. This wood has multiple uses in the wood industry in addition to domestically. The economic value of wood is basically dependent on its structure and chemical components. An understanding of the structure constituting wood and the different components of wood is necessary to assess the productivity and the quality of wood that is produced from a timber-yielding tree. Carbon, Nitrogen, and other chemical components constitute some of the important constituents of wood which differ in different concentrations among the diverse timber yielding trees. An understanding of the chemical constituents of wood is of profound interest to the ecologists for quantifying, analyzing, and interpretation of the causes and effects of some of the plant functional traits among the co-existing plant species (Chave et al., 2009; Moles et al., 2004; Westoby and Wright, 2006). An increase in the anthropogenic activities and uses of wood for a variety of purposes in the industry as fuel, timber, domestic purposes, etc., caused the utmost damage to the forests (Watson et al., 2018).

It is well recognized that trees have the inherent ability to take up carbon dioxide gas released into the atmosphere from different sources and by the capture of renewable bioenergy (solar energy) through leaves store this as wood carbon in the wood by the process of photosynthesis. Among the different trees, we come across large variations in the stored wood carbon contents. This is mostly accredited to the genetic variability in the tree species in their capacity to fix carbon in the process of carbon sequestration through leaves. This process is both physiological and physico-chemical as it involves the absorption of the atmospheric carbon dioxide and its storage in the biomass for longer durations. In other sense, it implies that carbon sequestration is a key process responsible for reversing the carbon dioxide accumulation in the atmosphere. Various researches have indicated that carbon dioxide is stored as wood carbon. Besides wood, carbon is also stored in other parts of the tree in the form of bioenergy. It is predicted that forests act as one of the sinks that store massive amounts of carbon.

The study outcomes showed that the forest cover in the US can accumulate 90% of the US carbon and act as a major sink of carbon. These forests may capture approximately 10% of the carbon dioxide emissions released from different sources in the US. According to Robert Canis, harvesting of trees may decrease the carbon emissions added to the atmosphere through the process of decomposition, and specified that in case of the mature forests, if harvesting of trees is increased it might serve as one of the mitigation options adopted for reduction of the carbon load in the atmosphere. Thus, harvesting not only promotes the wood as fossil fuel substitute can also reduce the carbon losses arising from the decomposition of parts of mature trees.

Very few studies are undertaken on wood carbon content. In Mexico, few studies have been carried out to assess the carbon content variability in the aboveground biomass (Rodríguez-Laguna et al., 2008; Jíménez-Pérez et al., 2013).

In this respect, Jiménez Pérez et al. (2013) analyzed for carbon content on per unit of dry weight (DW)-based biomass in the components of the aboveground biomass viz., stem, branches, bark, and leaves of the representative species *Pinus pseudostrobus, Juniperus flaccida, Quercus laceyi, Quercus rysophylla, Quercus canbyi* and *Arbutus xalapensis* of the pine-oak forest ecosystem of the Sierra Madre Oriental. Among these, *Juniperus flaccida* (51.18%) had the highest carbon concentration, while *Q. rysophylla* had the lowest (47. 98%). Further, it was observed that among the various components analyzed the *Arbutus xalapensis* (55.05%) leaves had the maximum carbon concentration, while the bark of *Quercus laceyi* had the lowermost carbon concentration (43.65%). There were significant differences in the average carbon concentration among the group of species. The conifer species showed an average carbon concentration of 50.76%, while the broadleaf species showed 48.85%. Apart from the variability in the average carbon concentration by a group of species, significant variations were also observed between the different components within a species group. The bark of conifers exhibited the presence of the highest carbon concentration (51.91%) than the bark of broad-leaved species which exhibited the lowest carbon concentration of 45.75%.

Thomas and Malczewski (2007) have undertaken a few studies on the interspecific variability and volatile wood carbon content of 14 native tree species in Eastern China. The results revealed highly significant variation among species in the wood C content that ranged from 48.4% to 51.0%. The volatile C fraction averaging 2.2% though was negligible, exhibited variations among the species. Research carried out in the species in their earlier studies has indicated that the conifer species had the highest wood C content of 50.80% than the wood C content (49.50%), in the hardwood of angiosperm species. Further, the variations were observed in wood carbon density (gC/cm^3) also. Very high inter-specific variations were observed in wood carbon density mainly due to the wood specific gravity difference. The results on wood C content published in North America demonstrated that the global mean value of wood carbon is 47.05% in dried wood, but little study is available on volatile C content of wood.

Zeng (2008), suggests that for mitigation of global climate change, it is required to develop strategies to uphold the atmospheric CO_2 concentration below the dangerous levels. It is estimated that a sustainable long-term carbon sequestration potential for wood burial is 10.5 GtC y^{-1}, and presently it is about 65 GtC y^{-1}. It is seen on the world's forest floors in the form of coarse woody debris that are suitable for burial. It was assessed that the tropical forests had a greater potential of sequestration (4.2 GtC y^{-1}), followed by temperate (3.7 GtC y^{-1}) and boreal forests (2.1 GtC y^{-1}). Therefore, burying of wood has other benefits as minimization of CO_2 source from deforestation, extension of the lifetime of reforestation carbon sink and reduction of fire danger.

All across the globe, it is of a major concern that continued expansion of terrestrial human footprints is leading to a faster decline of the native forest cover. There, might not be even a trace of native forest leftover free from human activities. It evidenced that at least the remaining intact forest that is existing on the earth, may help in combating the global environmental issues that have arisen due to the degradation of forests. As the forest cover is showing a rapid decline, it is very essential to preserve and uphold the traces of biodiversity, indigenous culture, human health, water provision, and carbon sequestration and storage. Therefore, the urgent priority before the countries across the globe for halting the biodiversity crisis, climatic changes or to attain the sustained goals or other global efforts undertaken to combat these problems is the maintenance and restoration of the lost intact forest cover (Watson et al., 2018).

Studies have been carried out on carbon fixation and nitrogen content in leaves of more than 40 woody species of Tamaulipan Thorn Scrub, Northeastern Mexico (Maiti et al., 2015). They reported a large variability among the 40 woody species in the carbon fixation (carbon sequestration). For the presence of high carbon and nitrogen content few woody trees and shrubs were selected. Some of the selected species with high carbon content were *Eugenia caryophyllata* (51.66%), *Litsea glauscensens* (51.54%), *Rhus virens* (50.35%), *Gochantia hypoleuca* (49.86%), *Pinus arizonica* (49.32%), *Eryobotrya japonica* (47.98%), *Tecoma stans* (47.79%), and *Rosmarinus officinalis* (47.77%). The research studies have indicated that because of the potentiality of these species to incorporate more amount of wood carbon into their biomass there is a possibility that these species if planted in the carbon polluted areas might reduce the carbon dioxide load in those areas and can be recommended for plantations across large areas.

A few species were also selected for the presence of nitrogen content. These include *Mimosa malacophylla* (8.44%), *Capsicum annuum* (6.84%), *Moringa oleifer* (6.25%), *Azadirachta indica* (5.85%), *Eruca sativa* (5.46%), *Rosmarinus officinalis* (5.40%), and *Mentha piperita* (5.40%). The above-selected species could also act as good sources of nitrogen too required for the maintenance of good health in animals. Apart from these certain species were selected with a high C/N ratio such *as Arbutus xalapensis* (26.94%), *Eryngium heterophyllum* (24.29%), *Rhus virens* (22.52%), and *Croton suaveolens* (20.16%). This may be related to the high production of secondary metabolites and antioxidants (Gonzalez Rodriguez et al., 2015).

Specific strategies were adopted to mitigate carbon emissions through forestry activities. Among these, the four major forest management strategies include the increase in the forest area by massive reforestation; increasing the carbon density of the existing forest stands both at the stand level and at landscape; increasing the usage of more number of forest products for replacing the emissions from the fossil fuels; reductions in the carbon emissions which aroused from increased rates of deforestation and forest degradation (Canadell and Raupach, 2008).

Although, increasing the carbon density of the existing stands and landscape is recommended as one of the mitigation strategies to decrease the carbon emissions, enactment of this requires the identification, selection, and planting of those trees which exhibit a faster growth habit and efficient capacities in the carbon fixation. However, it is well known to us that the forest cover and carbon dioxide absorption from the atmosphere can be increased by plantation of any kind of tree, the plantation of genetically modified tree specimens may be a suitable alternative, as these species have the ability to grow at a faster rate than the normal tree species.

In the context of the rate of deforestation (Canadell and Raupach, 2008), has shown that 13 billion meters squares of tropical regions are prone to deforestation annually. It is expected, these tropical region forests have a greater potential in carbon capture and emphasized that for stabilization of the global climate the deforestation rates in these tropical regions have to be reduced by 50% by 2050. In recent years, there is an increase in abandoned farmland and an increase in urbanization and intensive agriculture on farmlands.

A study was commenced by Kennedy (2009) to predict the wood density and carbon-nitrogen content in tropical agroforestry in Western Kenya using infra-red spectroscopy. He proposed that infrared spectroscopy coupled with chemometrics multivariate techniques is a fast and non-destructive cheap technique to get the fast and more reliable results. It was assessed that the measured carbon range was 40–52% (mean 48%), while IR predicted 44–51% (mean 48%) in the NIR region and 46–51% (mean 48%) in the MIR region. Measured nitrogen range was 0.09–0.48% (mean 0.28%), while IR predicted 0.18–0.47% (mean 0.24%) in NIR region and 0.18–0.38% (mean 0.24%) in MIR region. Interactions between densities with tree species and tree parts also revealed a significant effect (0.57 for all the parameters). This suggests that the large variations within species cannot be predicted using IR alone. NIR region gave better predictions than MIR. Their research findings have specified that the prediction performance was insufficient to recommend Infrared Spectroscopy as the practical method for direct determination of wood density and carbon content across species when different percentages were used.

Martin et al. (2015) in their studies on evaluation of the variations in the carbon and nitrogen contents among the major woody tissue types in temperate trees has mentioned that the quantification of the variations in the chemical traits of wood is very essential for the determination of forest biogeochemical budgets and models. They carried out their research studies on the analysis of wood carbon (C) and nitrogen (N) concentrations in 17 temperate tree species across five woody tissue types: sapwood, heartwood, small branches, coarse roots, and bark. Their analyses were also corrected for losses of volatile C. Their results have unveiled that significant variations existed for both C and N contents among the tissue types. These significant differences for C and N were observed in the bark of the species and appear to be the general pattern in the bark of all the species analyzed. Among the nonbark tissue types, the bivariate correlations among sapwood, heartwood, small branches, and coarse roots were found to be highly significant and positive for wood C ($r = 0.88$–0.98) and N ($r = 0.66$–0.95) concentrations. They proposed that for assessment and modeling of forest-level C dynamics the intraspecific variation in C across the tissue types is less important than the interspecific variation. In contrast, it was observed that the variances in N among tissue types were more and also appeared to be more important for incorporation into forest-level nutrient assessments and models. They suggest that, with the exception of bark, wood chemical trait values derived from stem wood can be used to accurately represent whole-tree trait values in models of forest C and N stocks and fluxes, at least for temperate species.

Recently, Varonin et al. (2017) has undertaken the isotope composition of carbon and nitrogen in tissues and organs of *Betula pendula*. They have identified the ratios of $^{13}C/^{12}C$ and $^{15}N/^{14}N$ isotopes in different parts and organs of drooping birch (*Betula pendula* Roth) in preforest-steppe and pine-birch forests of the Middle Urals by mass spectrometry. The data were analyzed and interpreted from the perspective of biochemical processes of carbon and nitrogen metabolism in the leaf, cambial tissue, trunk wood, branches, roots, and in the soil. It is concluded that the lighter isotopic composition of carbon is characteristic for the leaves, trunk, cambium and wood.

Gilson et al. (2014) undertook a very interesting study on seasonal changes in carbon and nitrogen compound concentrations in a *Quercus petraea* chronosequence. It is stated that forest productivity declines with tree age. This decline may be accredited to the variations in the metabolic functions, the resource availability, and/or changes in the resource allocation (between growth, reproduction, and storage) with tree age *in situ*. Their research results reflect a general pattern of carbon and nitrogen function at all tree ages, showing carbon reserve remobilization at budburst for growth, followed by carbon reserve formation during the leafy season and carbon reserve use during winter for maintenance. The variation in nitrogen compounds concentrations has shown less amplitude than carbon compounds. Storage as proteins takes place at later stages. It is mainly dependent on the leaf nitrogen remobilization and root uptake in autumn. Further, they observed the variations between tree age groups, mainly in the loss of carbon storage function of fine and medium-sized roots with tree aging. Moreover, the pattern of carbon compound accumulation in branches of their research findings supports the assumption of a preferential allocation of carbon towards growth until the end of wood formation in juvenile trees, at the expense of the replenishment of carbon stores, while mature trees start allocating carbon to storage right after budburst. The outcomes have demonstrated that at the key phenological stages, the physiological and developmental functions differ with tree age. Further, the environmental conditions also have an influence in sessile oaks for the variations in the carbon and nitrogen concentrations.

Sakai et al. (2012) estimated the wood density and carbon and nitrogen concentrations in deadwood of *Chamaecyparis obtusa* and *Cryptomeria japonica*. They commented that estimation of carbon (C) and nitrogen (N) stocks in deadwood in forests nationwide is needed to understand the large-scale C and N cycling. To do so, one requires the estimated values of wood density and C and N concentrations. It was assessed that wood densities decreased from 386 to 188 kg m^{-3} for *C. obtusa* and from 334 to 188 kg m^{-3} for *C. japonica* in decay classes but the variation in wood density increased with decay class, and the coefficient of variance increased from 13.9% to 46.4% for *C. obtusa* and from 15.2% to 48.1% for *C. japonica*. The N concentrations increased from 1.04 to 4.40 g kg^{-1} for *C. obtusa* and from 1.11 to 2.97 g kg^{-1} for *C. japonica* in decay classes 1–4. The variation in N concentration increased with decay class, and the coefficient of variance increased from 51.9% to 76.7% for *C. obtusa* and from 50.3% to 70.4% for *C. japonica*. Log diameter and region contributed to variations in wood density and N concentration in decay classes 1 and 2 for *C. obtusa* and *C. japonica*. However, no relationship was detected between regional climates and the two parameters. In contrast, C concentrations ranged from 507 to 535 g kg^{-1} and were stable with much lower coefficients of variance throughout the decay classes for both tree species. Thus, they recommend that the same C concentration can be adapted for all decay classes of both tree species.

From the perspective of the above discussions, the study was focused to determine the carbon contents in wood of 33 woody species at Linares, Northeastern Mexico.

5.2 MATERIALS AND METHODS

The study was undertaken at the experimental station of Facultad de Ciencias Forestales, Universidad Autonoma de Nuevo Leon, located in the municipality of Linares at an elevation of 350 m.

Wood of 33 woody shrubs and trees were collected. These were dried in oven at 60°C for 7–10 days until fully dried. After the complete drying, each sample of wood of each species is powdered separately using a Thomas Wiley mill. A 2.0 mg of the sample was weighed in an AD 6000 Perkin-Elmer balance in a vial of tin, bent perfectly. This was placed in CHONS analyzer Perkin Elmer Model 2400 for determining carbon and nitrogen. Carbon and nitrogen contents (% dry mass basis) were carried out in 0.020 g of milled dried wood tissue by using a CHN analyzer (Perkin Elemer, model 2400) (Figure 5.1).

[Procure the wood samples and dry them in oven at 60°C for 7–10 days until fully dried]

↓

[Powder separately each dried wood sample of each species using a Thomas Wiley mill]

↓

[Weigh 2.0 mg of the sample in an AD 6000 Perkin-Elmer balance in a vial of tin, bent perfectly]

↓

[Place the sample in CHN analyzer Perkin Elmer Model 2400 for determining carbon and nitrogen]

↓

[Calculate the carbon and nitrogen contents (% dry mass basis) in 0.020 g of milled dried wood tissue by using a CHN analyzer]

FIGURE 5.1   Flow diagram for the carbon and nitrogen contents determination in wood samples.

5.3   RESULTS AND CONCLUSIONS

Table 5.1 shows carbon and nitrogen contents in woods of 33 woody trees and shrubs, taken from the publication of Rodriguez et al., (2018) which is under review.

TABLE 5.1 Carbon and nitrogen contents in woods of 33 woody trees and shrubs

Scientific Name	%C	sd	%N	sd
Acacia berlandieri Benth.	51.00	1.69	1.07	0.03
Acacia farnesiana (L) Willd.	37.14	0.41	1.36	0.04
Acacia gregii var. wrightii Benth.	48.60	1.98	0.97	0.04
Acacia rigidula Benth.	48.23	0.01	0.85	0.00
Acacia schaffneri (S. Watson)	40.22	0.31	0.96	0.02
Amyris madrensis (S. Watson)	42.00	0.03	0.99	0.15
Amyris texana (Buckley) P. Wilson	49.70	1.18	1.18	0.07
Berberis chochoco Schlecht.	50.58	2.02	1.20	0.03
Bernardia myricifolia (G. Scheele) (S. Watson)	51.12	0.28	1.27	0.04
Caesalpinia mexicana A. Gray	45.73	2.12	0.99	0.06
Celtis laevigata Willd.	44.79	2.35	0.87	0.00
Celtis pallida Torr	46.39	1.83	1.15	0.03
Condalia hookeri M.C. Johnst.	43.26	2.67	0.71	0.04
Cordia boissieri A.DC.	38.55	0.38	1.26	0.02
Croton suaveolens Torr.	44.08	0.45	1.23	0.02
Diospyros palmeri Eastw.	43.86	0.78	1.01	0.00
Diospyros texana Scheele.	39.87	1.39	1.64	0.02
Ebenopsis ebano (Berland.) Barneby & J.W. Grimes.	45.90	1.82	1.97	0.01
Ehretia anacua I.M. Johnst	46.02	1.23	1.44	0.05
Eysenhardtia texana Scheele	48.17	1.58	1.31	0.06
Forestiera angustifolia Torr.	41.81	0.06	0.49	0.06
Guaiacum angustifolium Engelm.	47.53	1.03	1.52	0.03
Gymnosperma glutinosum (Spreng.) Less.	47.22	2.83	0.49	0.07
Havardia pallens (Benth.) Britton & Rose	50.36	0.74	0.77	0.01
Helietta parvifolia (A. Gray) Benth.	41.24	0.83	1.03	0.04
Karwinskia humboldtiana (Schult.) Zucc.	39.33	0.67	1.28	0.12
Lantana macropoda Torr.			0.55	0.02
Leucaena leucocephala (J. de Lamark) H.C. de Wit	48.86	0.74	1.06	0.00
Leucophyllum frutescens (Berland.) I.M. Johnst.	50.84	0.17	0.66	0.02
Parkinsonia aculeata L.			1.29	0.06
Parkinsonia texana (A. Gray) S. Watson	43.44	1.40	1.36	0.15
Prosopis laevigata (Humb. & Bonpl. Ex Willd.) M.C. Johnst.	47.63	0.45	1.29	0.13
Quercus virginiana P. Miller	47.99	2.79	0.71	0.02
Salix lasiolepis Benth.	42.42	1.03	0.56	0.00
Sargentia gregii S. Watson	48.14	2.14	1.23	0.02
Sideroxylon celastrinum Kunth. T.D. Penn.	46.92	2.33	1.75	0.02
Zanthoxylum fagara (L) Sarg.			1.37	0.03

It is observed that carbon content varied from 37 to 51 and nitrogen content from 0.55 to 1.97%. On the basis of the data mentioned, we selected woody trees and shrubs with high wood carbon content mentioned below.

Scientific Name	%C
Bernardia myricifolia (G. Scheele) S. Watson	51.12
Acacia berlandieri Benth.	51.00
Leucophyllum frutescens (Berland.) I.M. Johnst.	50.84
Berberis chochoco Schlecht.	50.58
Havardia pallens (Benth.) Britton & Rose	50.36
Amyris texana (Buckley) P. Wilson	49.70
Leucaena leucocephala (J. de Lamark) H.C. de Wit	48.86
Acacia gregii var. wrightii Benth.	48.60
Acacia rigidula Benth.	48.23
Eysenhardtia texana Scheele	48.17
Sargentia gregii S. Watson	48.14
Quercus virginiana P. Miller	47.99
Prosopis laevigata (Humb. & Bonpl. Ex Willd.) M.C. Johnst.	47.63
Guaiacum angustifolium Engelm.	47.53
Gymnosperma glutinosum (Spreng.) Less.	47.22
Sideroxylon celastrinum Kunth. T.D. Penn.	46.92
Celtis pallida Torr	46.39
Ehretia anacua I.M. Johnst.	46.02
Ebenopsis ebano (Berland.) Barneby & J.W. Grimes.	45.90
Caesalpinia mexicana A. Gray.	45.73
Celtis laevigata Willd.	44.79
Croton suaveolens Torr.	44.08
Diospyros palmeri Eastw.	43.86
Parkinsonia texana (A. Gray) S. Watson	43.44
Condalia hookeri M.C. Johnst.	43.26
Salix lasiolepis Benth.	42.42
Amyris madrensis S. Watson.	42.00
Forestiera angustifolia Torr.	41.81
Helietta parvifolia (A. Gray) Benth.	41.24
Acacia schaffneri (S. Watson)	40.22

We selected the five species with a very high value of wood C; (%) such as *Bernardia myricifolia* (51.12); *Acacia berlandieri* (51.00); *Leucophyllum frutescens* (50.84); *Berberis chochoco* (50.56); *Havardia pallens* (50.36).

It is highly recommended that the five selected species having 50% or more wood C mentioned above may be planted in cities, factories, sports, parks polluted with high carbon to reduce carbon from atmosphere. Besides, these could be recommended for incorporation in agroforestry for higher productivity of crops and timber. The leguminous woody species such as *Acacia* and *Leucophyllum* could improve soil fertility by nitrogen fixation capacity.

KEYWORDS

- *Arbutus xalapensis*
- **carbon dioxide**
- **chemometrics**
- **interspecific variability**
- *Juniperus flaccida*
- *Leucophyllum frutescens*

REFERENCES

Canadell, J. G., & Raupach, M. R., (2008). Managing forests for climate change. *Science, 320*, 1456–1457.

Chave, J., Coomes, D., Jansen, S., Lewis, S. L., Swenson, N. G., & Zanne, A. E., (2009). Towards a worldwide wood economics spectrum. *Ecol. Lett., 12*, 351–366.

Cherney, D. J. R., (2000). Characterization of forages by chemical analysis. In: Givens, D. I., Owen, E., Axford, R. F. E., & Omed, H. M., (eds.), *Forage Evaluation in Ruminant Nutrition* (pp. 281–300). CAB International, Wallingford, UK.

David Atkins, James Byler, Ladd Livingston, Paul Rogers, & Dayle Bennett (1999). Forest Health Protection Report 99-4. Missoula, MT: U.S. Department of Agriculture, Forest Service, Northern Region. 42 p.

Gilson, A., Barthes, L., Delpierre, N., Dufrêne, E., Fresneau, C., & Bazot, S., (2014). Seasonal changes in carbon and nitrogen compound concentrations in a *Quercus petraea* chronosequence. *Tree Physiol., 34*, 716–729.

Gonzalez Rodriguez, H., Maiti, R. K., Valencia Narvaez, R. I., & Sarkar N. C., (2015). Carbon and nitrogen content in leaf tissue of different plant species, northeastern Mexico. *Intern. J. Biores. Stress Manage.*, *6*, 113–116.

Jiménez Pérez, J., Treviño Garza, E. J., & Yerena Yamallel, J. I., (2013). Carbon concentration in pine-oak forest species of the Sierra Madre Oriental. *Rev. Mex. Ciencias For.*, *4*, 50–61.

Kennedy, O. O., (2009). *Prediction of wood density and carbon-nitrogen content in tropical agroforestry in Western Kenya using infrared spectroscopy.* University of Nairobi, School of Physical Sciences, Department of Chemistry. Thesis, Master of Science. 110 p.

Maiti, R., Rodriguez, H. G., & Kumari, Ch. A., (2015). Trees and shrubs with high carbon fixation/concentration. *Forest Res. Open Access, S1*, 003. doi: 10.4172/2168-9776. S1-003

Martin, A. R., Gezahegn, S., & Thomas, S. C., (2015). Variation in carbon and nitrogen concentration among major woody tissue types in temperate trees. *Can. J. For. Res.*, *45*, 744–757. dx.doi.org/10.1139/cjfr-2015-0024.

Moles, A. T., Falster, D. S., Leishman, M. R., & Westoby, M., (2004). Small-seeded species produce more seeds per square meter of canopy per year, but not per individual per lifetime. *J. Ecol.*, *92*, 384–396.

Peters, R. L., (1990). Effects of global warming on forests. *Forest Ecol. Manage.*, *35*, 13–33.

Rodríguez-Laguna, R., Jiménez-Pérez, J., Meza-Rangel, J., Aguirre-Calderón, O., & Razo-Zarate, R., (2008). Carbon content in a semi-deciduous tropical forest at Reserva de la Biosfera El Cielo, Tamaulipas, Mexico. *Rev. Latinoamer. Rec. Natur.*, *4*, 215–222.

Sakai, Y., Ugawa, S., Ishizuka, S., Takahashi, M., & Takenaka, C., (2012). Wood density and carbon and nitrogen concentrations in dead wood of *Chamaecyparis* obtusa and *Cryptomeria japonica*. *Soil Sci. Plant Nutr.*, *58*, 526–537.

Thomas, S. C., & Malczewski, G. (2007). Wood carbon content of tree species in Eastern China: interspecific variability and the importance of the volatile fraction. *J. Environ. Manage.*, *85*, 659–662.

Voronin, P. Y., Mukhin, V. A., Velivetskaya, T. A., Ignatev, A. V., & Kuznetsov, V., (2017). Isotope composition of carbon and nitrogen in tissues and organs of *Betula pendula*. *Russ. J. Plant Physiol.*, *64*(2), 184–189.

Watson, J. E. M., Evans, T., & Venter, O. et al., (2018). The exceptional value of intact forest ecosystems. *Nat. Ecol. Evol.*, *2*, 599–610.

Westoby, M., & Wright, I. J., (2006). Land-plant ecology on the basis of functional traits. *Trends Ecol. Evol.*, *21*, 261–268.

Westoby, M., Falster, D. S., Moles, A. T., Vesk, P. A., & Wright, I. J., (2002). Plant ecological strategies: Some leading dimensions of variation between species. *Ann. Rev. Ecol. Syst.*, *33*, 125–159.

Zeng, N., (2008). Carbon sequestration via wood burial. *Carbon Bal. Manage.*, *3*, 1–12.

CHAPTER 6

Leaf/Wood Chemical Composition

6.1 INTRODUCTION

Wood is the key raw material of the wood industry used for the manufacture of furniture, other domestically useful articles, and other products. The wood quality, which is mostly dependent on the anatomical structure and chemical composition, differ with the type of timber species. These variations also determine the use of wood for different purposes.

The main chemical constituents of wood are cellulose, hemicelluloses, lignin, and other extractable materials. The lignin is present in the cell wall of wood (20–30%). It is mostly impregnated in the cell wall of wood, and imparts rigidity of the same. Lignin also acts as an obstacle against the cell wall degrading enzymes. The lignin is distinguishable from other cellulosic materials. In wood, it is always associated to cellulose, but do not occur in the same association in other cellulosic materials. Therefore, some cellulose is also found in a pure condition, for example, in cotton (Ortuño, 1993). Lignin is a tridimensional aromatic polymer in which phenyl propane units are repeated with different types of bonds (ether or C-C) between the monomers.

Some research studies were undertaken on the chemical composition of woods in relation to the wood quality and its utilization. Pettersen (1984) reported the chemical composition of various woods. His research studies dealt with the overall chemical composition of wood, the methods of analysis, the structure of hemicelluloses components and the degree of polymerization of carbohydrates. These components were analyzed and the chemical composition of woods of several countries both in the United States and other countries were compiled. Further, he also studied the individual sugar composition (as glucan, xylan, galactan, arabinan, and mannan), uronic anhydride, acetyl, lignin, and ash components include:

cellulose (Cross and Bevan, holo-, and alpha-), lignin, pentosans, and ash. He also reported the solubilities of wood in 1% sodium hydroxide, hot water, ethanol/ benzene, and ether. He studied common temperate-zone woods and reported the individual sugar composition (as glucan, xylan, galactan, arabinan, and mannan), uronic anhydride; a wood contains two major chemical components in wood: lignin (18–35%) and carbohydrate (65–75%). Both are complex, polymeric materials. In addition, minor amounts of extraneous materials, mostly in the form of organic extractives and inorganic minerals (ash), are also present in wood (usually 4–10%). Overall, wood is composed of an elemental composition of about 50% carbon, 6% hydrogen, 44% oxygen, and trace amounts of several metal ions cetyl, lignin, and ash.

Bárcenas-Pazos and Dávalos-Sotelo (1999) undertook a review on the importance of lignin on the contraction of woods. Based on literature, it is established that the shrinkage of wood can be attributed partially to the presence of the lignin and its content in the wood. Temperate hardwood species, obtained both from Mexico and the United States showed a greater shrinkage percentage than tropical hardwoods and softwoods from both countries. The three-dimensional stiffness of lignin was more than that of the other chemical constituents of the cell wall. It was found that lignin is mostly related with the limit of the movements due to variations in the moisture contents owing to its presence of low hygroscopicity. Regression analyses were carried out to find out the effect of specific gravity and lignin content. These revealed that though both these traits have a marked importance, the main variable is the specific gravity. On the other hand, it is found that lignin also has a significant effect. They proposed that it is essential to carry out experimental studies on the influence of these variables on dimensional changes, together with other important variables, such as extractives and ray volume.

Woody species in particular and all the plants comprise the major components of elements as carbon, hydrogen, oxygen, and nitrogen. They also contain minor quantities of calcium, potassium, and magnesium. The elements carbon, hydrogen and oxygen are combined in different propor-tions to form the organic components of wood viz., the cellulose, hemi-cellulose, and lignin and also some pectin compounds (Ortuño, 1993). The polysaccharides and lignin form the major components of cell wall. Celluloses and hemicelluloses constitute the major components of the polysaccharides.

The main component of the cell wall of all higher plants and (40–45%) of wood fibers is cellulose. Cellulose is constituted by β-D-glucose in form of pyranose linked together by 1-4-glycosidic bonds with the formation of cellobiose residues. The hemicellulose is also found in the cell wall. Often it is found to be associated with cellulose in the cell wall. This is formed by pentose and hexose distinct from glucose (mannose, xylose, glucose, galactose, and arabinose), linked together with a polymerization grade from 100 to 200. A large variation is seen in the species regarding the chemical structure and the composition of the constituents. All the hemicelluloses are insoluble in H_2O. However, they can be dissolved in strong alkalis and easily hydrolyzed by acids. The ability to have a higher solubility and susceptibility to hydrolysis by hemicelluloses than the cellulose molecules is due to its amorphic structure and low molecular weight (Ortuño, 1993). The cellulosic fraction-cellulose and hemicellulose of wood may be separated in its components, depending on its solubility in NaOH at 17.5%, according to their grade of polymerization (Ortuño, 1993).

The pectins or pectic substances are also the hydrates of carbon. These also form the cell wall of young cells. The method of extraction of lignin from wood is variable, often it becomes difficult to separate the lignin component from the wood fractions. The molecular weight of the separated product may vary between 1000 and 20000 g/mol (Lu and Ralph, 2010). The presence of the high lignins, aromatic and phenolic compounds contents imparts dark color to the wood. Lignins are easily oxidized. They are comparatively stable in aqueous acidic minerals. Lignins are soluble in aqueous bases and hot bisulfite.

There are also a series of extractible compounds such as gums, resins, fats, alkaloids, and also tannins in the wood in varying chemical compositions. These compounds can be extracted from wood by cold or hot water, or organic solvents such as alcohol, benzene, acetone, or ether. The proportion of these substances though varies from 1 to 10%, some tropical species may even contain up to 20% of the same. The inorganic compounds are not soluble in the mentioned solvents, but sometimes are included among the extracts (Ortuño, 1993). The extraneous matters are constituted by substances which can be separated by extraction by non-reactive dissolvents, protein residues of the protoplasm of the growing cell and mineral components, some of them very difficult to remove.

Bertaud and Holmbom (2004) in their analysis of the wood components distribution using microscale analytical technique along the radial cross-section of stem revealed that lignin is present to a large extent in the heartwood. Further, it was found that heartwood had more lignin and less amounts of cellulose than that of the softwood. However, the total hemicellulose content was found to be similar along the radial direction in both the heartwood and the softwood. In contrast, the transition zone between heartwood and sapwood contained specific chemical components with less lignin and lipophilic extractives. The sugar unit's distribution in the hemicellulose showed significant differences. Latewood contained galacto glucomannan while the earlywood had less pectin.

Jones et al. (2006) undertook a study of the estimation of the wood chemical compositions of the radial strips of wood with the diffuse reflectance near-infrared spectroscopy method. Radial longitudinal face strips of seventeen wood strips of *P. taeda* were obtained from seven sites. They obtained the NIR spectra from a 12.5 mm section of radial strip representing the predetermined positions of the juvenile wood (close to pith), transition wood (zone between juvenile and mature wood), and mature wood (close to bark). By means of the standard analytical chemistry methods cellulose, hemicellulose, lignin (acid-soluble and insoluble), arabinan, galactan, glucan, mannan, and xylan contents were determined in these radial strips. By utilizing the data obtained from the NIR spectra, wood chemistry, and partial least squares (PLS) regression the calibration curves were developed. Though the best results were obtained for cellulose, glucan, xylan, and lignin, the relationships were found to be variable. They observed that there was high prediction errors consequence of the diverse origins of the samples that were tested.

Ray et al. (2012) determined the physical and chemical properties of the wood of the mangroves. They also determined their annual carbon sequestration ability. They determined the growth rate on the basis of the data obtained from the diameter, breast height, wood density, etc. Similarly, by using FTIR spectroscopy and thermogravimetry, the assessment of carbon sequestration has shown that there was an increase in the rate of carbon sequestration rate with density. This has varied between 0.088 and 0.171 g C kg^{-1} AGB s^{-1}. Maximum carbon sequestration was observed in *Avicennia marina,* and the minimum was noticed in *Bruguiera gymnorrhiza.* There were changes in FTIR bands at 4000–2500 cm^{-1} and 1700–800 cm^{-1}. These were found to show positive correlations with the cellulose and lignin contents of the wood. Further, it was observed that cellulose and lignin content in the wood of mangroves is in the range of 0.21 and

1.75. The fuel value index of mangrove wood ranged between 985–3922. The fuel index was found to be negatively related to the decomposition temperature and density. They observed that the seasonal variation of temperature and CO_2 not only brings differences in the wood density but also has an effect on the wood chemical properties.

Chandrasekaran et al. (2012) studied the chemical composition wood chip and wood pellet samples manufactured in the United States and Canada analyzed for their energy and chemical properties and compared to German standards for pellet quality. The calorific value, moisture content, and ash content of the samples were determined according to the American Society for Testing and Materials (ASTM) methods. Using ASTM methods sulfate and chloride samples were prepared and analyzed by ion chromatography (IC). With inductively coupled plasma mass spectrometry (ICP-MS) the elemental compositions of the ashed wood samples were determined. Mercury was measured by direct wood samples analysis. The sample characteristics distributions, such as heating value, ash content, moisture content, ions, and heavy elements, are presented. Major ash-forming elements were Ca, K, Al, Mg, and Fe.

Chandrasekaran et al. (2012) reported that an analysis of chemical composition of 132 wood pellet and 23 wood chip samples of those which were manufactured in Canada and USA, had an average chemical contents of wood, (carbon 45–50%; 6.0–6.5%; Oxygen 38–42%; Nitrogen 0.1–0.5%; Sulphur max 0.05). In the following the main chemical components of some wood were; species; cellulose range from 39 to 45%; hemicellulose 28 to 32%; lignin 22 to 31%.

Baharoğlu et al. (2013) investigated the wood anatomical and chemical structures effects on the quality properties of particleboard containing a different mixture of wood species. They concluded that the anatomical and chemical structures were effective on all of the properties of particleboards. Panels made from the particles including more amount of pinewood showed the highest mechanical strength properties and lowest thickness swelling values. Cellulose, hemicellulose, and lignin contents, acidity, and solubility values (in hot-cold water, dilute alkali and alcohol benzene) of wood significantly affected all of the properties of particleboards. The physical and mechanical properties of particleboards showed statistical differences related to the length, thickness, and number of the cells and fiber.

A study was conducted by Pawlicka and Waliszewska (2013) on chemical composition of selected exotic wood species resultant from the area of Africa by determining the contents of: cellulose, lignin, holocellulose, pentosans, and substances soluble in organic solvents, in 1% NaOH solution, in cold and hot water, in addition to the contents of mineral substances. The performed investigations revealed that concentrations of the determined constituents in individual wood species varied despite similar site and climatic conditions shown below.

The results show that Koto species contained cellulose – 43%, hemicellulose – 72%, lignin – 22%, while Mahogany species contained cellulose – 43%; hemicellulose – 60%; lignin – 30%.

6.2   MATERIALS AND METHODS

A study has been carried out in Forest Science Faculty, Universidad Autonoma de Nuevo Leon, Mexico, on the variability of wood density of ten woody species and its possible relation to wood chemical composition and wood anatomy. The authors studied the relationship of wood density with wood chemical composition and wood anatomy. In the present study, a large variation in wood density was also observed. It was found that among nine species studied, there exists a large variation in wood density (0.51 to 1.09), and few wood chemical compositions such as % carbon (37.14 to 44.07), nitrogen (9.18 to 19.22), sulfur (31.45 to 33.82), lignin (15.28 to 24.35), hemicellulose (19.94 to 27.36), and % cellulose (33.69 to 45.92). In general, though there was no clear association between wood density and other chemical composition of wood, it is observed that the species having moderate to high wood density contained >40% carbon, >30% sulfur, and >40% cellulose and more or less 20% lignin. It seems that carbon, sulfur, cellulose, and lignin content contribute to greater density. The wood fiber cell wall lignification seems to be associated with wood density (Maiti et al., 2016).

Recently, Scharnweber et al. (2016) studied within- and between-tree variations of wood-chemistry measured by X-ray fluorescence. They investigated the degree of synchronism of elemental time series within and between trees of one coniferous (*Pinus sylvestris* L.) and one broadleaf (*Castanea sativa* Mill.) species growing in conventionally managed forests without direct pollution sources in their habitats. They used micro-X-ray fluorescence (µXRF) analysis to establish time series of relative concentrations of multiple elements (Mg, Al, P, Cl, K, Ca, Cr, Mn, Fe, and Ni) at different stem heights and stem exposures. In both species, for most elements, a common long-term (decadal) trend was observed but only little coherence was observed in the high-frequency domain (inter-annual variations). The pine had a stronger common signal than for chestnut. Many elements in pine show a concentration gradient with higher values. The present study was commenced on the chemical composition of 37 woody shrubs and trees in Northeast Mexico.

6.2.1 PREPARATIONS OF SAMPLES AND TECHNIQUES USED FOR CHEMICAL COMPOSITION

Wood samples of each species were ground in a Thomas Willey mill (Thomas Scientific Apparatus, Model 3383) with N60 (1 mm × 1 mm) mesh, these were sieved and stored in labeled plastic containers. Samples by triplicate were subjected to chemical analysis. The neutral detergent fiber (NDF), acid detergent fiber (ADF), and acid detergent lignin (ADL) contents were determined by methods defined by Van Soest et al. (1991) (ANKOM procedure). Hemicellulose (NDF-ADF) and cellulose (ADF-lignin) were acquired by difference. The estimation of each component was done in five replications.

6.3 RESULTS AND CONCLUSIONS

According to Rodríguez et al. (not published), wood is of great commercial importance in the wood industry. Its quality depends on its chemical composition. The present study was carried out on the variability in the chemical composition of 37 woody trees and shrubs of Tamaulipan Thorn Scrub, Northeastern Mexico. The results show large variability among the species in wood chemical composition such as NDF, ADF (digestible detergent fiber), lignin, cellulose, and hemicellulose. In these aspects, the maximum of NDF (94.8%) is observed in *Celtis pallida, Parkinsonia aculeata,* and *Guaiacum angustifolium.* Much more ADF was found in three species: *Celtis pallida, Parkinsonia aculeata,* and *Lantana macropoda.* With regard to lignin, *Sideroxylon celastrina, Ebenopsis ebano, Ehretia anacua, Amyris texana, Leucophyllum frutescens, Cordia boissieri,* and *Condalia hookeri* have high lignin content (over 24%). However, maximum of cellulose is observed in *Parkinsonia aculeata*. *Celtis pallida* and *Lantana macropoda* have also high levels of cellulose. Similarly, the maximum of cellulose is observed in *Bernardia myricifolia*. Close to, the maximum content of hemicellulose was revealed in *Celtis laevigata*. Regarding fiber, a maximum of fiber is observed in *Celtis pallida*. Close to, the maximum content of fiber was revealed in *Parkinsonia aculeata* and *Guaiacum angustifolium*. The variations in chemical compositions could be related to the quality determination and utility of timbers of different woody species (Rodriguez et al., 2016).

6.4   RESEARCH NEEDS

Research needs to be directed to determine the quality and use of wood of species containing a high percentage of chemical components such as cellulose, lignin, etc. for the fabrication of furniture, paper pulps, etc.

KEYWORDS

- cellulose
- chemical composition
- extractable materials
- hemicellulose
- lignin

REFERENCES

Baharoğlu, M., Nemli, G., Sarı, B., Birtürk, T., & Bardak, S., (2013). Effects of anatomical and chemical properties of wood on the quality of particleboard. *Composites: Part B: Engineering, 52*, 282–285.

Bárcenas-Pazos, G. M., & Dávalos-Sotelo, R., (1999). Importancia de la lignina en las contracciones de la madera: revisión bibliográfica. *Madera y Bosques, 5*, 13–26.

Bertaud, F., & Holmbom, B. R., (2004). Chemical components of early and heartwood in Norway spruce heartwood and sapwood and transition zone. *Wood Sci. Technol., 38*, 245–256.

Chandrasekaran, S. R., Hopke, P- K., Rector, L., Allen, G., & Lin, L., (2012). Chemical composition of wood chips and wood pellets. *Energy Fuels, 26*, 4932–4937.

David, S. Y. W. (2016). *Wood Chemistry: Fundamentals and Application.* Professor Department of Forestry, NCHU.

Jones, P. D., Schimleck, L. R., Peter, G. F. et al. (2006). Nondestructive estimation of wood chemical composition of sections of radial wood strips by diffuse reflectance near infrared spectroscopy. Wood Sci. Technol., 40, 709–720.

Lu, F., & Ralph, J., (2010). Lignin. *In:* Cereal Straw as a Resource for Sustainable Biomaterials and Biofuels. S. Run-Cang (ed.). Elsevier. Amsterdam, The Netherlands. pp. 169–207.

Maiti, R., Rodriguez, H. G., & Kumari, A., (2016). Wood density of ten native trees and shrubs and its possible relation with a few wood chemical compositions. *Am. J. Plant Sci., 7,* 1192–1197.

Ortuño, A. V., (1993). Introduction to industrial chemistry. Editorial Reverté. Barcelona, España. p. 656.

Pawlicka, A., & Waliszewska, B., (2011). Chemical composition of selected species of exotic wood derived from the region of Africa. *Acta Sci. Pol. Silv. Colendar. Rat. Ind. Lignar., 10,* 37–41.

Pettersen, R. C., (1984). The chemical composition of wood. Chapter 2: The chemistry of softwood. *Advances in Chemistry, 207,* 57–126.

Ray, R., Majumder, N., Chowdhury, C., & Jana, T. K., (2012). Wood chemistry and density: An analog for a response to the change of carbon sequestration in mangroves. *Carbohydrate Polymers, 90,* 102–108.

Rodriguez, H. G., Maiti, R., Kumari, A., & Sarkar, N. C., (2016). Variability in wood density and wood fiber characterization of woody species and their possible utility in northeastern Mexico. *Am. J. Plant Sci., 7,* 1139–1150.

Scharnweber, T., Hevia, A., Buras, A., van der Maaten, E., & Wilmking, M., (2016). Common trends in elements? Within- and between-tree variations of wood-chemistry measured by x-ray fluorescence: A dendrochemical study. *Sci. Total Environ.,* 566–567, 1245–1253.

Van Soest, P. J., Robertson J. B., & Lewis B. A., (1991). Methods for dietary, neutral detergent fiber, and non starch polysaccharides in relation to animal nutrition. Symposium: carbohydrate methodology, metabolism, and nutritional implications in dairy cattle. *J. Dairy Sci., 74,* 3583–3597.

PART II

LABORATORY TECHNIQUES OF PHYSIOLOGY AND BIOCHEMISTRY

INTRODUCTION

This part describes various basic techniques used in the laboratory for various biophysical, physiological, and biochemical components.

CHAPTER 7

Revision of Terminology and Laboratory Techniques

The main objective of this chapter is to familiarize the students and techni-
cians with some techniques of the laboratory. The key points discussed are
the details of the techniques utilized during the semester. These concepts
and techniques, of course, are not new, but previous experiences have
demonstrated that the majority of the students do not remember all these
techniques. One should know the basic principles of laboratory techniques
and the methods of application before planning out any research program.

Plant physiology is based on the concepts of physics and chemistry,
and also of the use of terms and the systems of measurement of parameters
used in these sciences. Plant physiology is a science, and therefore, should
be described and presented in precise terms following its basic principles.
The students should understand the basic principles, and functions of plant
physiology in plant growth and development. In this session, many students
will get familiarized with the terms used and materials used in the laboratory, however, the students should not confuse with the basic knowledge.

7.1 SOLUTIONS

All living beings depend on water. The protoplasm of a plant cell contains
various cell organelles dispersed in water present in the cytoplasm, and
therefore, all the materials of cells are transported with water. Besides,
almost all biological reactions occur in aqueous solutions. The physical,
chemical, and electrical properties of the solutions and the dispersion of
the materials in water form the basis for chemistry and physics of the living
material. Hence, it is very important to understand clearly the properties
of the solutions.

7.1.1 UNITS OF CONCENTRATIONS

1. **Molar Solution:** One molar solution (M) contains molecular weight in gram (MW) for mol of one substance per liter of solution.

M = mols of solute/liter of mol. solution = grams of solute/MW (g)

2. **Normal Solution:** One normal solution (N) contains one equivalent weight in grams per liter of solution. The equivalent weight of one substance depends on the nature of the reaction for the solution used. In the reactions of acid-base, the equivalent weight of acid or base is that quantity of material produced or consumed by one molecular weight of protons. In the reactions of oxidation-reduction, this corresponds to the quantity of the material which produces or consumes one equivalent of electrons (one molecular weight of atoms of hydrogen).

N = equivalent weight of solute/liter of solution

Equivalent = grams of solute/equivalent weight, in grams

3. **Molal Solution:** One molal solution (m) contains molecular weight in grams of one substance (Note the differences between molal solution and molar solution). The molality is used in situations, where for reasons physical or chemical reactions, it is desirable to express the reasons of molecules of solvent, for example, used in discussions on osmotic pressure.

where m = mols of solute/kilogram of solvent structures or the molecular weight are not known.

There are three forms where percentage solution may be used and each one should be mentioned clearly.

i.   **Weight Per Unit of Weight (w/w):** For example, one solution of 20% (w/w) of sugar in water signifies 20 grams of sugar + 80 grams of water or in other words 20 g of sugar per 100 g of solution.

ii. **Weight Per Unit of Volume (p/v):** For example, one solution of 20% (w/v) of sugar in water signifies 20 grams of sugar and add water until 100 ml are reached or in other words, 20 grams of sugar per each 100 ml of solution.

% (w/v) = grams of solute /100 ml of solution.

iii. **Volume Per Unit of Volume (v/v):** In general, it is used for the solution of liquid in liquid. For example, One solution 80% (v/v) of alcohol in one solution in water signifies 80 ml of alcohol in 100 ml of a solution, i.e., 80 ml alcohol and 20 ml of water.

For the facility and convenience of the students, we try to use dilutions volume-volume. This is one part of the concentrated solution, and ten parts of that solution, or is one part of the concentrated solution, and nine parts of solvent to make a final volume of 10 parts. We may write one dilution 1 in 10 in various manners: 1 in 10, or 1:10 or 1 10. This signifies that one part of one solution has been diluted to make a final volume of 10 parts (more nine parts). The final concentration of a solution is a one-tenth part of the original concentrated solution (1/10). One dilution 2:7 signifies that 2 parts of the concentrated solution have been diluted in one final volume of 7 parts (2 + 5). The final concentration of the solution (or solute) is two/seventh (2/7) of the original concentration. Inversely, the original solution of concentrated gives 7/2 or 3.5 times more concentrated than the final solution.

7.1.2 DILUTION IN SERIES

Series of dilutions are used to obtain one series of solutions in which the concentrations of solute change by some constant of proportion or ratio. This is done regularly to stabilize one curve of calibration, needed especially in bioassays of growth regulators, pigments, or calculation of nutrients. For example, it is desirable to measure the effects of the herbicide in one particular process or want to test different concentrations, or in different magnitudes. We have to realize a series of dilutions, in which the solutions differ from one to another by one solution of 1 a 10. If 9 ml of 5 concentrations are required, the higher concentration comes from 1 mg/ml. For these operations, we need five test tubes. In the first tube, we have 10 ml of concentrated stock solution. In the second tube, we should have one 1 ml of first and 9 ml of dilution. In the third tube, we should have 1 ml of second and 9 ml of dilution and similarly successive dilutions.

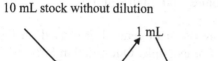

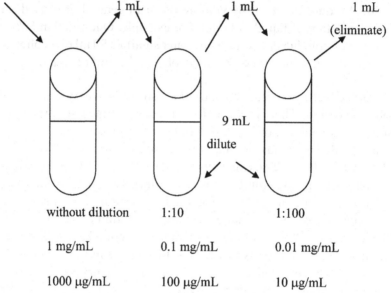

FIGURE 7.1   Series of dilution.

7.1.3 CALCULATIONS

With an objective to change one dilution to others, there are four aspects that a student should know. These are concentrations and volume of the desired solution, and concentration and volume of the original solution. In fact, only three of these variables are given, these are presented in the following manner:

$$\text{Volume} \times \text{Concentration} = \text{Volume} \times \text{Concentration}$$
$$\text{(initial)} \qquad\qquad\qquad \text{(final)}$$

This allows to calculate an unknown variable. This relation is valid where the quantity of solute is the same in both sides of the equation and is exactly in the same volume.

➤ *Example:*

Given a solution of 1 g per ml

Obtaining 10 ml of 0.3 g/ml

Volume (stock) × 1 g/ml = 10 ml × 0.3 g/ml

Volume = 3 ml

To prepare 10 ml of 0.3 g/ml of one stock solution of 1 g/ml, take 3 ml of stock solution and dilute it to final volume 10 (or add 7 ml of solvent i.e., distilled water).

➤ *Example:*

Preparation of 0.3 M NaCl of 25 ml from a stock solution of 5M NaCl.

Volume to be taken × known standard concentration = Volume required × required concentration of solution

x ml × 5 = 0.3 × 25

x ml = 0.3 × 25/5 = 1.5

Take 1.5 ml of 5M stock solution of NaCl and make up the volume to 25 ml with distilled water or 1.5 ml of 5M NaCl and 23.5 ml of distilled water.

7.1.4 PIPETTES

The pipettes are used to transfer liquids exactly. Each one who works in a science laboratory should know properly about the pipettes for liquid transfers, how to identify pipettes and should have the capacity to collect correct pipettes required for a particular work.

There exist diverse classes of pipettes that are normally used: serological (graduated up to tip); Mohr or measurement (not graduated up to the tip); volumetric (not graduated) and microliters or lambda.

All volumetric crystals are calibrated, either for contents of pipette (PC) or liberation (PL) one specific volume of liquids. Take only the glass pieces which have been calibrated, i.e., the pipettes for liberation (PL) and consider the quantity of liquid adhered on the surface of pipettes after draining out the liquids. There is no necessity to take out residues. Those pieces of crystals which have been calibrated such as PC include the total volume of liquids which are adhered on the surface of pipettes after emptying. These pieces should be washed nicely, because in general in these pipettes the solution is prepared with the purpose to transfer the quantity of liquid required. The measuring and serological pipettes are calibrated such as PL; on the other hand, the volumetric pipettes or pipettes of transfer might be of type PL or PC.

The volumetric pipettes are used to contain a large volume of solution. While drawing out a solution by using a pipette, the solution in the pipette is maintained vertically and the liquid is extracted by suction (either with mouth or with the help of a suction syringe and the outside is cleaned nicely). The liquid is then liberated completely. The contents are, then, emptied inside the receptor vessels until the flow is detained. It will take about 10–20 seconds so that the liquid of the wall of the pipette is collected up to the tip by rotary movement, maintaining the tip away from the receptor vessel. Do not blow the remaining liquid out for blowing away. The volumetric pipettes used specially in biochemical work are calibrated for blowing away. These Ostwald pipettes are marked with opaque rings of 1/8" breadth located near the extreme of suction. Some designated pipettes blow by two opaque rings near the tie of the tip. When volume is required immediately, one measuring or serological pipette should be used. You should be assured also of the size of the appropriate pipette. Do not use, for example, one pipette of 10 ml to release one volume of 0.5 ml. The error is associated with the transfer in general 1% of calibrated volume. One pipette of 10 ml may be an error of 1%. This liberates ± 0.1 ml of volume indicated. One pipette of 1 ml may commit the error of 1% to liberate or release ± 0.01 ml of volume indicated. If one pipette of 10 ml was used to release, the error may be ± 0.1 ml or 10%, is while it is one pipette of 1 ml is used to liberate 1 ml, the error would be 1%. The liquids should be drained up to line zero. The pipette is then dried and kept up to the liquid goes down to line zero. The last drop is removed by touching the tip of the pipette in the receptor vessel. The quantity of liquid required is then liberated in the receptor vessel to assure to include the last drop. The serological pipette is calibrated to blow out when all the contents are taken out.

After completion of the usage of the pipettes, they should not be retained in the containers or chemical solutions from which a required sample was taken out. To eliminate the accidents and contamination of reactive, the used pipette should always be washed thoroughly and placed in a pipette stand.

The instructor of the laboratory will demonstrate the correct method of management of pipettes and suction knob.

7.1.5 CONCENTRATION OF HYDROGEN ION (H⁺) AND pH

In physiology, it is essential to know the concentration of H^+ present in a solution. Many of these processes are influenced by changes in the concentration of protons. For convenience, one logarithmic scale is used to express $[H^+]$. The pH is defined as the negative logarithm of the activity H^+, which is frequently expressed in molar concentration.

This is expressed mathematically as follows:

$$pH = -\log [H^+], \text{ when } [H^+] \text{ is expressed in moles/liter.}$$

In this scale, one small number signifies that $[H^+]$ is large, and it is said that the solution is acid. One large number signifies that $[H^+]$ is small, and it is said that the solution is basic. In general, the acids are defined as substances that form H^+ in water and the bases are defined as substances that combine and neutralize H^+. This is expressed as follows:

$$K_w \text{ (ionic product of water)} = [H^+] [OH^-] = 1.0 \times 10^{-14}$$

This complies when the concentration of H^+ and OH^- are equal (10^{-7} M), then the pH is 7 and the solution is defined as neutral.

With this definition, 0.01 M HCl is 2 because $[H^+]$ is 0.01 or 10^{-2} M; and $-\log (10^{-2})$ is 2. Simultaneously, the pH of 0.01 M of NaOH is 12, because $[OH^-]$ is 0.01 or 10^{-2} M. This is expressed as $[H^+] \times [OH^-]$ should be equal to 10^{-14}, $[H^+]$ should be equal to a 10^{-12}, and $-\log (10^{-12})$ is 12.

Really speaking, the pH's are slightly different than those values calculated owing to presence no acids and bases. Besides, the additional deviations occur in the case of weak acids and bases, so that these do not dissociate completely although these require equal base or acids. As a result, one complete neutralization occurs. For example, 1 M HCl has a pH of 0.1 in place of 0.0 which may be calculated. Simultaneously 1 M Acetic acid has a pH of 2.37 which uniquely 0.4% of molecules dissociated.

There are two cases that students need to remember in the calculation of pH. The first is that a change of one unit in pH (for example, 4 vs. 5) represents one change of 10 times of 10 of [H (or bases), especially the organics may contribute (to neutralize) more than that one H^+ (for example, malic or citric acid).

7.1.6 OBJECTIVE

To familiarize the students with the analytical process involved in the preparation of solutions, mathematical calculations and the correct use of materials such as pipettes in different preparations.

7.2 METHODOLOGY

7.2.1 PIPETTES

Realize the following activities.

1. Describe various pipettes which might be used during the semester in various experiments.
2. Realize the following transformations:
    i.   1 ml with serological pipette;
    ii. 1 ml with measuring pipettes;
    iii. 5 ml with one volumetric pipette;
    iv. The instructor will ask the student to take one and ask the student to justify the selection of the pipette.

7.2.2 BALANCES

The instructor demonstrates the use of the operation of a balance and asks the students to record the following in an observation notebook.

1. Unknown weight of a sample of seed lot;
2. Weighing 5 g of seeds of one species given.

7.2.3 SOLUTIONS

Prepare 50 ml of the following solutions

1. 0.5 M NaCl;
2. 0.1 N Citric acid;
3. 1% sucrose (w/v);
4. 0.005 M NaCl using the initial solution A along with the stock solution, all these measurements and calculations should be noted in the observation notebook.

7.2.4 CALCULATIONS OF pH

1. pH of a solution of 0.05 M of H_2SO_4 (Assume the complete dissociation).
2. pH of a solution of 0.10 N of H_2SO_4.
3. pH of a solution of 0.001 M of HCl.
4. pH of a solution of 0.001 N of HCl.

7.3 REPORT THE FOLLOWING

1. How can you prepare 2 ml of each of the following dilutions: Be specific:

    i.   Without dilutions;
    ii.   1:2;
    iii. 1:4;
    iv.   1:8;
    v.   1:16.

2. What is the name of this process?
3. If the stock solution is 32 mg/ml, what is the concentration of each dilution?
4. How will you prepare the following solution?

    i.   100 ml of 0.001 M H_2SO_4 from a stock solution of 0.1 M?
    ii.   50 ml of 0.02 M NaCl?
    iii. 10 ml of 10^{-5} M of IAA from a stock solution of 10^{-3} M?
    iv.   50 ml of 1 mM of 2,4-D from one stock solution of 0.1 M?

5. Distinguish between the volumetric pipette and serological pipette.
6. What is the difference between crude mass and weight?

KEYWORDS

- cytoplasm
- microliters
- molecular weight
- negative logarithm
- pipettes for liberation
- serological pipette

CHAPTER 8

Diffusion and Osmosis

8.1 INTRODUCTION

The atmosphere is constituted by various types of gases which take place in air, the average of the kinetic energy of each species (i.e., oxygen) being essentially identical. This reveals, although when the kinetic energy of the molecule of oxygen varies significantly without giving importance to its history of collusion. The average kinetic energy of one gas is equal to $\frac{1}{2}mv^2$, where m is the mass (g) and v is its velocity (cm s^{-1}). Therefore, different classes of molecules of gases have identical kinetic energy but with different masses, but should have a constant temperature. The small molecules possess greater velocity, and if there is, no impediments and the molecules move through stomatas to outside the leaves rapidly than those with greater weight.

The terminology is known as the root of the average square velocity of gas, with no exact value but very near to the average of its velocity. This value may be calculated from the following Eq. (1) (*Graham's law*):

$$\mu = \sqrt{\frac{3RT}{M}} \qquad (1)$$

where, R is the constant of an ideal gas is expressed in 8.314×10^7 erg K^{-1} mol^{-1}, T is the absolute temperature ($273.16 + °C$), and M is the molecular weight of gas (g mol^{-1}).

From the Eq. (1), we may calculate the average approximate velocity of various velocities of gases at different temperatures. Under conditions of normal atmospheric pressure, whatever molecule in the air which in general travels very near before its course is changed by the collision of other molecules of the same of different classes. Similarly, from the Eq. (1), we can demonstrate that the average approximate velocities of different classes of gases, at the same temperature, are related with respect of the other by the following Eq. (2):

$$\frac{\mu_1}{\mu_2} = \frac{\sqrt{M_2}}{\sqrt{M_1}}$$

(2)

Under specific conditions, the time required for two gases to diffuse through of very small orifice are inversely proportional to their average velocities and therefore are directly proportional with root squares of their molecular weight. This proportion is uniquely exact when the area of orifice is very small than the length which is the process of diffusion occurring (i.e., when the distance traveled is sufficiently small to minimize collisions with other molecules). All the molecules of gases neither have the same kinetic energy nor displace in the same velocity. This may be that under these conditions may interchange energy, and the velocity also changes continuously the direction of one molecule. In whatever samples of one gas, there occur a large number of molecules in the same form so that the molecule velocities are distributed at the interval of distinct velocities. However, at the normal atmospheric pressure, they travel freely in the medium (distance among collisions) is short between 150 to 400 times of the diameters of the particles. With such high velocities and travel very briefly among collisions of each molecule is enormous in the order of thousands of millions per second. Table 8.1 illustrates that the change in temperature from 0 to 30°C, which covers large intervals of temperature for proper functions of the life, the average velocities of particles increase is only around 5%.

8.2　OBJECTIVE

To estimate the relative velocities of NH_3 and HCl traveling in opposite directions up to the center of the crystal tube. These gases will form NH_4Cl when they collide.

8.3　MATERIALS

1. One tube of glass hollow crystal, of 1 cm of diameter × 60 cm of length.

2. One-piece to support the tube and one support of iron with a base of 12.5 cm of breadth and 15 cm of length and one sweep of 60 cm in length a (Figure 8.1).
3. Two wads of cotton to close the crystal tube.
4. Two lids of cork (o rubber stoppers) to close the crystal tube and two pins.
5. One pair of pencils or forceps and a ruler.
6. Two beakers of 100 ml which contain HCl and NH_4OH.

TABLE 8.1 Some Molecular Values of Three Gases

Property	H_2	O_2	CO_2
Molecular weight of gas (Da)	2.01	32.0	44.0
Average velocity at 0°C, (m s⁻¹)	1696	425	362
Average velocity at 30°C, (m s⁻¹)	1787	448	382
Average velocity at 100°C, (m s⁻¹)	1682	497	424
Free travel in the medium (nm) among collisions with other molecules at 0°C and one atmospheric pressure	112	63	39
Number of collisions of each molecule per second, in thousands of millions of miles (1 x 10⁹), at 0°C and one atmospheric pressure	15.1	6.8	9.4
Diameter of each molecule (nm)	0.272	0.364	0.462
Number of molecules (x 10⁻¹⁹), at 0°C and one atmospheric pressure, in 1 cm³	2.70	2.71	2.72

8.4 METHODOLOGY

In the chamber of vapor extraction, place crystal tube horizontally in one extension rod which is subjected to supporting iron rod (Figure 8.1). With one forceps insert one cotton plug at the extreme end of the tube of lower diameter using cork. Repeat this in other cork. Using forceps moisten cotton with 12 N HCl and other with 15 N NH_4OH. Eliminate the excess of acid with absorbent/tissue paper. Insert the lids of cork with cotton plug concurrently in the opposite ends of the crystal tube (Figure 8.1).

Observe the tube carefully until one white ring is formed inside the tube, near the extreme end which contains cotton boll moistened with acid. Later measure the distance of the white ring.

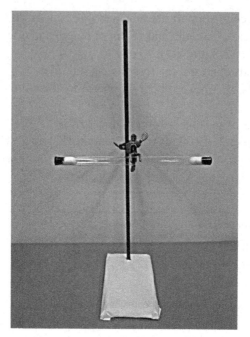

FIGURE 8.1   Diagram to illustrate the diffusion of a gas.

8.5   REPORT

8.5.1   QUESTIONS

1. From definitions of ergs and dynes, calculate root squares of average velocity in gaseous molecules of O_2, H_2O, and CO_2 at 27°C and 15°C?. Form a table to facilitate the presentation of the results.
2. If gases NH_3 (1) and HCl (2) are diffused rapidly as molecules of question 1, why was the ring of NH_4Cl not detected in an instantaneous manner?

8.5.2   RESULTS

Distance of the diffusion of NH_4OH: ....

Distance of the diffusion of HCl: ....

Ratio, $\dfrac{1}{2}$ =

Molecular weight of HCl: ....

Molecular weight of NH_3: ....

$$\text{Ratio,} \quad \boxed{\dfrac{\sqrt{M_2}}{\sqrt{M_1}}} =$$

KEYWORDS

- **collisions**
- **forceps**
- **kinetic energy**
- **square velocity**
- **stomata**
- **velocities**

CHAPTER 9

Imbibition of Seeds

9.1 INTRODUCTION

The surface of hydrophylls (e.g., the colloids such as proteins, starch, and clays) adsorb water and the tenacity with which it occurs depends not only on the nature of surface, but also on the distance among surface and water molecules which adsorb: which is present on adsorbent surface and retained with great forces. The adsorption of water on the surface of hydrophylls is known as hydration or imbibition. The imbibition is a physical phenomenon and is independent of temperature. Besides, this process is not the function of the viability of seeds. Therefore, it does not depend on metabolism and may take place in the absence of oxygen.

The imbibition of water in a dry seed induces a sequence of chemical reactions which induce the germination (protrusion of radicle through seminal segments) and subsequent seedling development. It is well known that the physiologists define the germination process as an event which initiates with imbibition and terminates with the emergence of radicle (embryonic root or in some seeds the emergence through seed testa (Bewley, 1997). One seed may remain viable (living) but has no ability to germinate or grow for many reasons such as dormancy which needs to be broken under favorable condition or seed treatment. These may be classified or forced under extreme or internal conditions. One internal situation which is easy to understand is one embryon which has not reached morphological maturity for germination (e.g., in some members of Orchidaceae, Orobanchaceae, or genus *Ranunculus*). Only time permits for attaining maturity. The germination of the seeds of wild plants is frequently limited by this or other internal factors or only for the absence of moisture and or hot. So as to differentiate between these two different situations, the physiologists have utilized two terms: quiescence referring to the condition of one seed which cannot germinate only because when adequate external conditions are not available (for example when the seed is very dry or cold), and due to dormancy, because under these condition the seeds cannot germinate for internal condition (temperature, moisture, and atmosphere) are not adequate.

Plant physiologists identify four stages which take place during the germination process: (1) the hydration or imbibition during which water enters the embryon and hydrate proteins carbohydrates and other colloids, (2) the formation or activation of enzymes which increase metabolic activity, (3) elongation of the cells of radicle followed by the protrusion of radicle through testa (this is proper germination stage), and (4) the subsequent growth of the seedling. The layers cover the embryon (endosperm, seminal, and fruit cover) may interfere with the entry of water and oxygen or both, and may impede the emergence of radicle acting as mechanical barriers. In other seeds, it seems that it impedes the escape of inhibitors of embryo or these themselves contain inhibitors of germination.

9.2   OBJECTIVE

To demonstrate the imbibition process of seeds relative to osmotic effects.

9.3   METHODOLOGY

9.3.1   OSMOTIC EFFECTS ON SEED IMBIBITION

Prepare 100 ml of the following molar solutions of NaCl: 4.0, 2.0, 1.0, 0.5, 0.25, and 0). Put each solution in Erlenmeyer flask of 250 ml previously labeled. The osmotic potential (ψ) of these six solutions are: -197.97, -98.98, -49.49, -24.74, -12.37 and 0 bars, respectively. Carefully weigh in analytical balance exactly 10.0 g of seeds of the lot assigned and place the seeds in one beaker. Prepare the other five samples in the same manner which have been dried previously at 103°C. Register the weight of each sample in your observation notebook.

Place each group of 10.0 g per solution of NaCl prepared and seals with parafilm the mouth of flask to avoid evaporation. Maintain the flasks at a constant temperature of 25°C. After 2 days drain out completely the solution, dry, and weigh each seed sample again.

Put in graphs the results obtained in x- and y-axis. It represents the grams of water imbibed by one group of seeds (g H_2O g^{-1} seed), and (bars) of the solution as abscissa. Explain the results in the report) (Figure 9.1).

[Prepare 100 ml of the molar solutions of NaCl: 4.0, 2.0, 1.0, 0.5, 0.25, and 0]

↓

[Put each solution in Erlenmeyer flask of 250 ml]

↓

[Carefully weigh in analytical balance exactly 10.0 g of seeds of the lot assigned]

↓

[Place the weighed seeds in one beaker]

↓

[Prepare others 5 samples in the same manner which have been dried previously at 103°C]

↓

[Register weight of each sample in your observation notebook]

↓

[Place each group of 10.0 g per solution of NaCl prepared and seal with parafilm the mouth of flask]

↓

[Maintain the flasks at a constant temperature of 25°C]

↓

[After 2 days drain out completely the solution, dry, and weigh each seed sample again and plot the results on a graph paper]

FIGURE 9.1   Flow diagram for the study of osmotic effects on seed imbibition.

i.   How should be the form of the curve if one equivalent concentration of sucrose is used?

ii. Why do seeds imbibe water?

9.3.2   RATE OF IMBIBITION

Place 5.0 g of seeds of maize in one beaker with about 30 ml of water and mark the beaker as R-1, weigh the other 5 g and place in other beaker and mark the beaker as R-2. Repeat the same for the third beaker and label it as R-3. Allow that the seeds get wetted by stirring strongly. Remove all the seeds that are floating.

After wetting the seeds in water, remove them, dry, and weigh carefully, registering the weights in your observation notebook. Be careful to maintain observations separately of the seeds of lot R-1, R-2 and R-3 vessels with water. Remember the initial weight of seeds registered as: t = 0.

Repeat the stirring process and dry the seeds after 2, 4, 6, 10, 18, 24, 48, and 72 hours. Through graphs, illustrate the increase of weight in g, (axis of the ordinate) against time of imbibition, hours (abscissa). Table 9.1 helps to register your data. Remember that the increase of weight is equal to weight in a particular time given (t = x; where x = 2, 4, 6, 10, 18, 24, 48 and 72 h) showing lower weight at time t = 0 (Figure 9.2).

TABLE 9.1   Weight of Seeds (g) at Different Times of Imbibition (h) in Water

Time of imbibition	Replication			
(h)	R-1	R-2	R-3	Average
0				
2				
4				
6				
10				
18				
24				
48				
72				

[Place 5.0 g of seeds of maize in one beaker with about 30 ml of water
and mark the beaker as R-1]

↓

[Weigh other 5 g and place in other beaker and label
the beaker as R-2]

↓

[Repeat the same for the third beaker and label it as R-3. Allow that the seeds get
wetted by
stirring strongly and remove floating seeds]

↓

[After wetting the seeds in water, remove them, dry, and weigh
carefully, registering the weights in your observation notebook]

↓

[Prepare others 5 samples in the same manner which have been
dried previously at 103°C]

↓

[Register weight of each sample in your observation notebook]

↓

[Repeat the process of stirring and dry the seeds after
2, 4, 6, 10, 18, 24, 48, and 72 hours]

↓

[Through graphs illustrate the increase of weight in g,
(axis of the ordinate) against the time of imbibition, hours, (abscissa)]

FIGURE 9.2   Flow diagram for the study of the rate of seed imbibition.

9.4   REPORT

1. Compare data of each student team with the rest of the teams.
2. Interpret your results obtained from the graphs generated and compare them with different species of seeds.

KEYWORDS

- **colloids**
- **embryonic root**
- **analytical balance**
- **hydrophylls**
- **imbibition**
- **parafilm**

REFERENCES

Bewley, J. D., (1997). Seed germination and dormancy. *The Plant Cell, 9*, 1055–1066.

CHAPTER 10

Estimation of Water Potential in Potato Tuber from Changes in Weight and Volume

10.1 INTRODUCTION

The principle of this experiment is based to findout a solution of known water potential (in other words, osmotic potential) where plant tissue does not gain or lose water. This is demonstrated through changes of weight or volume of tissue. Thereby, we can select species susceptible or resistant to drought stress.

Species having higher water potential in a semiarid environment are considered as resistant to drought.

The pressure potential of whatever solution at any atmospheric pressure is conveniently zero. Similarly, the water potential of one sucrose solution is equal to osmotic potential. This is valid for all the solutions and tissue in equilibrium, but uniquely where the tissues do not gain or lose water from its contents or initial water potential, as in the sucrose solution.

The water potential values obtained by this technique are in general more negative than those obtained by other techniques (e.g., the method of Chardakov). This is owing to the fact that intercellular spaces and cell walls are filled with water where cell protoplasm is in osmotic equilibrium. This hydration allows to increase slightly the weight and volume of tissue which implicates, of course, small errors during the measurement of water potential. Therefore, special care needs to be taken while measuring water potential.

10.2 OBJECTIVE

To estimate the changes in the water potential of potato tissue and determine the changes in weight and volume after the equilibrium of the tissue with solutions of different osmotic potentials.

10.3   MATERIALS

1.   Two potato tubers are sufficient for the study;
2.   200 ml of 1.0 M of sucrose (20°C to 25°C);
3.   Two graduated measuring cylinders one of 25 ml and other 50 ml;
5.   One beaker of 150 ml;
6.   Five beakers of 250 ml;
7.   A cork borer of 4 or 5 mm diameter;
8.   One thermometer;
9.   One rectangular piece of wood with two separate knives of 30 mm;
10.   Moistened towel or filter paper;
11.   One analytical balance with a sensitive to less than one milligram.

10.4   METHODOLOGY

From solution of 1.0 M of sucrose prepare 100 ml of the following (M): 0.10, 0.20, 0.25, 0.30, and 0.35. Put each solution in the beaker previously labeled. Then, using the cork borer, obtain 15 cylinders of potato tissue from tubers (Figure 10.1). Cut each cylinder to 30 mm in length using cutters and be sure that the tuber tissue does not have the layer of epidermis. Then each cut cylinder is placed in a beaker of 150 ml and covered with paraffin to minimize evaporation from cut surface. When all segments are cut, 15, select three cylinders randomly at a time and weigh, as a group, in an analytical balance, to the nearest 0.01 g. Note down the weights in your notebook as described in Table 10.1. Measure and note the average of three lengths of potato cylinders to a precision of 0.5 mm, and then place in one of the solutions of sucrose. Repeat the determinations of weight and lengths of the different groups subsequently, each one in three sections. Place each group of cylinders in sucrose solutions of different concentrations. Cover the beakers with paraffin to reduce evaporation and maintain at least 120 minutes at a temperature of 20 to 25°C to reach osmotic equilibrium. After the incubation time, move one group of potato tissue and dry surface with paper towels or filter paper. Weigh each group of cylinders to a precision 0.01 g and measure its length corresponding to a precision of 0.5 mm. Note down the data in your observation notebook. Repeat the same for other groups of cylinders (Table 10.1).

TABLE 10.1 Distribution of Weights and Lengths (Cylinders) of Potato Tuber Tissue Exposed to Different Concentrations of Sucrose

Solution	Weight, g				Length, mm	
	Initial	Final	Increase (Δ)	%Δ	Initial	Final
0.10 M						
0.20 M						
0.25 M						
0.30 M						
0.35 M						

Develop a graph to illustrates the change (%) in potato tissue weight as a function of sucrose concentration (Figure 10.2). The increase in weight is shown above the zero line, while the decreases are presented below the zero line. Connect points with a line which you consider better fits your data. The point at which the fitted line crosses the sucrose concentration or zero change line, represents an estimation of tissue water potential very near to a solution that may have the same water potential as the potato tissue tuber. Figure 10.3 shows a sequence diagram for the procedures for this experiment. Obtain the osmotic potential of each solution by using the van't Hoff equation; $\Psi = -RTc$, where Ψ is the osmotic potential (MPa), R is the ideal gas constant (8.34×10^{-3} L MPa mol^{-1} K^{-1}), T is the temperature in Kelvin ($273.16 + 25°C$), and c is the molar concentration (mol L^{-1}) of solute (sucrose).

10.5 REPORT

10.5.1 QUESTIONS

1. What are your conclusions obtained from this experiment?
2. What are the main sources of error in this experiment?
3. What considerations are more important: that potato sections should have same inicital weight or should have same dimensions? Explain.

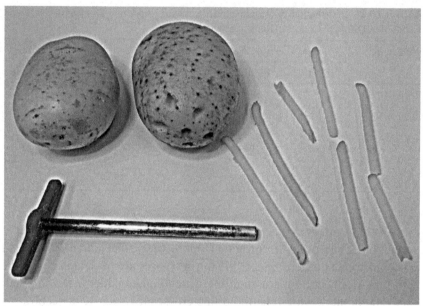

FIGURE 10.1   Processing to obtain the cylinders of the potato tuber tissue.

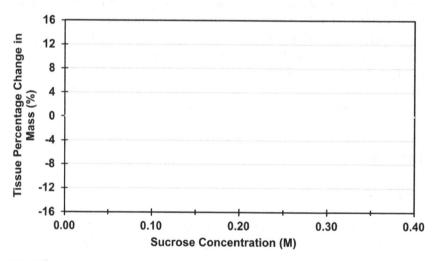

FIGURE 10.2   Effect of sucrose concentration on gain or loss of water in potato tuber tissue.

[From a solution of 1.0 M of sucrose, prepare 100 ml
of the following molar soluction: 0.10, 0.20, 0.25, 0.30, and 0.35]

↓

[Put each solution in beaker previously labeled]

↓

[By using a cork borer, cut 15 potato cylinders of 30 mm in length each]

↓

[Out of the 15 potato tissue sections, select randomly 3 cylinders a time and
weigh, as a group, using an analytical balance, obtain precisely 0.01 g]

↓

[Measure and note the average of three lengths of the potato pieces to a
precision of 0.5 mm, and then place in one of the solutions of sucrose]

↓

[Repeat the determinations of weight and lengths of the
different groups subsequently, each one in three sections]

↓

[Place each group of cylinders in sucrose solutions of
different concentrations]

↓

[Cover the beakers with paraffin to reduce evaporation and maintain at least 120
minutes at a temperature of 20 to 25°C to reach osmotic equilibrium and after
the incubation time, weigh the sample]

FIGURE 10.3   Flow diagram for estimation of water potential in potato tuber tissue from
the changes in weight and volume.

KEYWORDS

- Chardakov
- epidermis
- equilibrium
- incubation
- sucrose
- water potential

CHAPTER 11

Influence of pH on Coloration of Anthocyanin and Betalaine

11.1 INTRODUCTION

The flavonoids are compounds of 15 carbons which are found in the vegetable kingdom. There are more than 2,000 flavonoids classes have been identified from different plant tissue sources that contribute to the color of leaves, fruits, flowers, and seeds (Kumar and Pandey, 2013; Panche et al. 2016). The basic structure of the flavonoids (Figure 11.1) is commonly found to be modified in such a form which is present in more than double bonds so that the compounds absorb visible light.

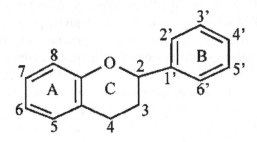

FIGURE 11.1 Basic structure of the flavonoids.

Flavonoids are based upon a fifteen-carbon skeleton structure consisting of two benzene rings (A and B as described in Figure 11.1) linked via a heterocyclic pyrane ring (C) (Kumar and Pandey, 2013). Flavonoids can be subdivided into different subgroups depending on the carbon of the C ring on which the B ring is attached and the degree of unsaturation and oxidation of the C ring. Flavonoids in which the B ring is linked in position 3 of the C ring are called isoflavones. Those in which the B ring is linked in position 4 are called neoflavonoids, while those in which the B ring is linked in position 2 can be further subdivided into several subgroups on the basis of the structural features of the C ring. These subgroups are: flavones, flavonols, flavanones, flavanonols, flavanols or catechins, anthocyanins and chalcones (Kumar and Pandey, 2013). The most of the flavonoids are accumulated in the central vacuole even though they are synthesized away from it. Three groups of flavonoids that are of greater significance in plant physiology are anthocyanins, flavonol, and flavons.

11.2 ANTHOCYANINS

The anthocyanins (from Greek *anthos*, flower, and *kyaneos/kyanous*, dark blue) are colored pigments, commonly found in red, blue, and purple flowers. They are also found in other plant structures, such as in few fruits, stems, leaves including roots. In general, the flavonoids are found confined in epidermal cells. The majority of flowers and fruits have their colors because of the presence of anthocyanins, although in some yellow flowers and tomato fruits are colored by the presence of carotenoids. The brilliant colors of the leaves in autumn are attributed, to a great extent, due to the accumulation of anthocyanins in cool days and high light intensity.

In higher plants there exist different types of anthocyanins and generally may be found in more than one type in a given plant tissue structure. Anthocyanins are present as glycosides, in general with one or two units of glucose or galactose in the central ring of the hydroxyl group, or in the hydroxyl group located in position 5 of the ring. When the sugars are excluded, from the molecules, these are known as anthocyanidin molecules.

The anthocyanidins are named in conformity to the specific plant species where they come from. The anthocyanins vary only in a number of hydroxyl units of ring B of the basic flavonoid structure.

The color of anthocyanins depends in the first place on substitute groups which are present in ring B. When the methyl groups are present, such as peonidin provoke one effect of redness (Figure 11.2). The majority of anthocyanins are reddish in acidic solution but it takes purple or blue with an increase in pH. Due to these properties and the presence of more anthocyanins, there exists a wide variation in the color of flowers.

Based on literature, various possible functions of anthocyanins have been discussed. One of their useful functions in flowers is the attraction of honeybees which carry pollen from one plant to other, which favors the pollination. Charles Darwin suggested that the attractive color of fruits serve to attract birds and animals which may eat fruits and disperse their seeds. The anthocyanins may also contribute in resistance to diseases, although evidences in favor of this are weak. Certainly, their abundance suggests some functions which have favored their evolutionary selection.

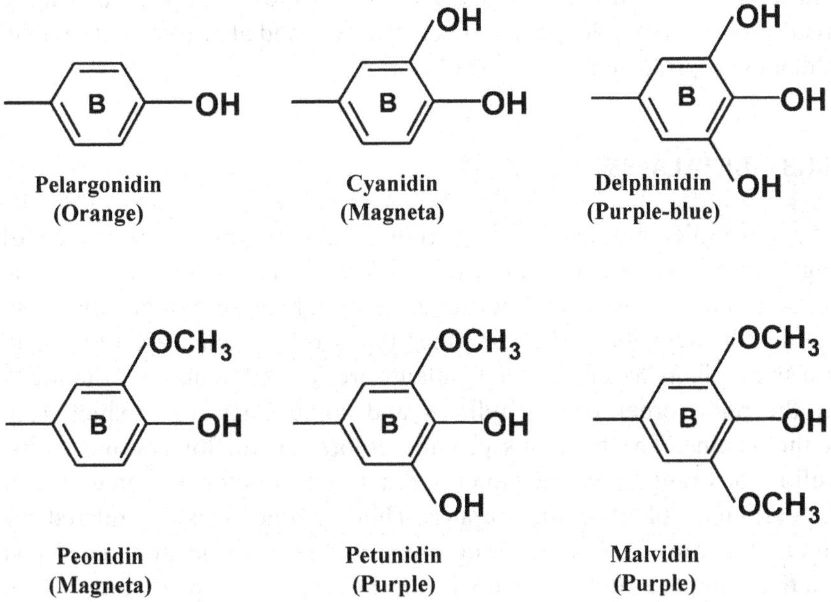

FIGURE 11.2   Variations in ring B owing to hydroxylation and methylation to form diverse anthocyanidins.

Flavonols and flavons are very much related with anthocyanins because they differ in a central ring structure which contains oxygen (Figure 11.3).

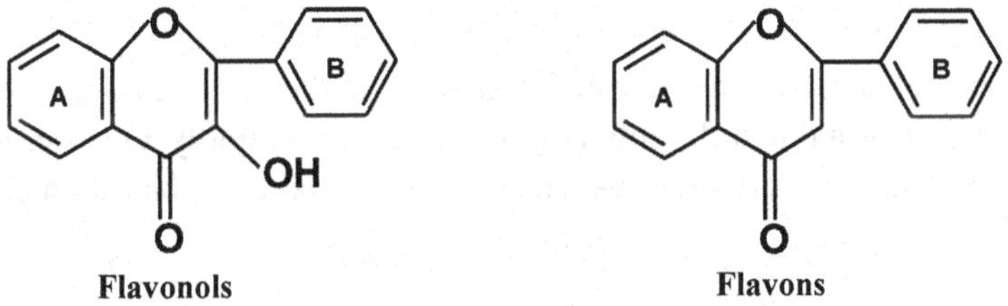

FIGURE 11.3   Basis skeleton structure of flavonols and flavons.

Most of the flavons and flavonols are colorless or extremely pale yellow, function as copigment substances (Iwashina, 2015) and as anthocyanins, frequently contribute to the color of the flower. Although some flavons and flavonols are colorless, they sill abosrb ultraviolet (UV) wavelengths with which modify the visible spectrum of radiation for bees and other insects that are attracted by the flowers containing these pigments. These molecules also are very common in leaves. It appears that these also serve as favorable factors of consumption and also absorb ultraviolet radiation as protection against UV rays.

11.3 BETALAINS

The red-violet pigment of beet root is a betacyanin, one group of pigmented betalains of red color, which was supposed for a long time to be associated with anthocyanins, although these contain nitrogen. Neither betacyanin (red), nor other types of betalains are present in the same plants together. The betalains are limited to about 10 families of the plant order Caryophyllales and some Cactaceae, which lack anthocyanins. The betalains provide coloration to flowers and fruits, yellow or orange-red and violet and also gives color, to some extent, to particular plant tissue organs. Their synthesis is stimulated by light. Betalains also contain sugar and one colored portion. The most studied member of this group is the betalain from beet root, which may hydrolyze with glucose and betanidin, one red pigment with the following structure (Figure 11.4).

Its function appears to be involved in pollinization in a comparable manner as anthocyanin does in other species: protection against insects a possible function.

Since 1960, it is known that these two pigments are chemically different. The anthocyanins, in general, have positive charges which move to negative pole during the process of electrophoresis, while the betacyanins commonly have one net negative charge which moves up to the positive pole. However, electrophoresis aid in differentiating these pigments rapidly. Other criteria to find these pigments are their response to differential color based on changes in pH.

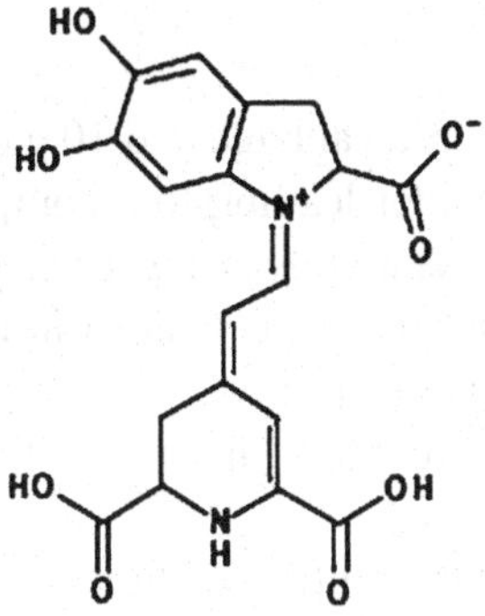

FIGURE 11.4 Basic structure of betainidin.

11.4 OBJECTIVE

To observe the response in the changes in anthocyanins and betalain exposed to change of pH.

11.5 MATERIALS

1.	10 g of red cabbage
2.	10 g of beetroot
3.	1 balance
4.	1 blender
5.	1 Buchner funnel
6.	2 Filter paper
7.	2 Sidearm flask of 500 ml
8.	1 Vacuum pump
9.	10 Test tubes of 15 ml
10.	1 rack
11.	5 ml of 0.1 N HCl
12.	5 ml of 0.01 N KOH
13.	5 ml of 0.1 N KOH
14.	2 tablets of solid NaOH or KOH
15.	15 ml of 1 N HCl
16.	1 Pasteur pipet
17.	4 Pipettes
18.	1 Erlenmeyer flask of 125 ml

11.6   METHODOLOGY

Liquefy separately 10 g of red cabbage and 10 g of beetroot with 200 ml of distilled water using a blender leading to a completely homogenous solution. Filter each solution separately using filter paper, a Buchner flask, a Sidearm flask, and vacuum pump. Transfer filtrate in an Erlenmeyer flask and remove solid material not filtrated.

Add 5 ml of solution of cabbage in each of five test tubes labeled from 1 to 5.

Repeat the same for the solution of beetroot but numbered from 6 to 10. Maintain the tube no 1 and 6 as controls (no treated or control).

Add 1.0 ml of HCl 0.1N in those test tubes nos. 2 and 7. Mix well and note in the notebook whatever changes of color occurs.

Add 0.5 ml of KOH 0.01N in test tube no. 3 and 8 and mix well the solution of cabbage change to violet, or, if too basic, blue. Note the color of both solutions. In the tubes of no. 4 and 9 add 1.0 ml of KOH 0.1N and note the changes of color.

Finally add one tablet of solid, KOH, or NaOH in the tubes no. 5 and no. 10. Anthocyanin change color to yellow. Observe these colors.

Later determine by addition of drops of HCl 0.1 N or HCl 1N for the tubes no. 5 and no. 10 and see if the changes of color are reversible (Figure 11.4).

11.7   REPORT

11.7.1   QUESTIONS

1. Describe the colors of the original filtrate of cabbage and beetroot in the notebook.
2. Do the pigments show the same changes of color on the addition of acid or a basic solution?

[Liquefy separately 10 g of red cabbage and 10 g of beetroot with 200 ml of
distilled water using a blender leading to a completely homogenous solution]

↓

[Filter each solution separately using filter paper in a Buchner flask,
one Sidearm flask and vacuum pump]

↓

[Transfer filtrate in an Erlenmeyer flask and remove solid material not filtrated]

↓

[Add 5 ml of solution of cabbage in each of five test tubes marked from 1 to 5]

↓

[Repeat the same for the solution of beetroot but numbered from 6 to 10
Maintain the tube no 1 and 6 as controls (no treated or control)]

↓

[Add 1.0 ml of HCl 0.1N in those test tubes no. 2 and 7]

↓

[Add 0.5 ml of KOH 0.01N in test tube no. 3 and 8 and mix well.
Note the color of both solutions]

↓

[In the tubes of no. 4 and 9 add 1.0 ml of KOH 0.1N and
note the changes of color]

↓

[Finally add one Tablet of solid, KOH or NaOH in the tubes
no. 5 and no. 10. Observe these colors]

↓

[Later determine by addition of drops of HCl 0.1 N or HCl 1N for the tubes no.
5 and no. 10 and see if the changes of color are reversible]

FIGURE 11.5 Flow diagram for the influence of pH on the coloration of anthocyanin
and betalain.

3. While analyzing the pH effects on the coloration of anthocyanins, do you consider that the difference of pH in plant tissue is corresponding to diverse variations observed in flowers which contain anthocyanin?
4. Was there reversible changes of color?

11.7.2 OBSERVATIONS

The observations described the changes in the color of the extracts of beetroot and cabbage in the conformity of exposure of these to an increase of HCl or KOH.

KEYWORDS

- **anthocyanins**
- **betalain**
- **Erlenmeyer flask**
- **Sidearm flask**
- **pigments**
- **pollinization**

REFERENCES

Iwashina, T., (2015). Contribution to flower colors of flavonoids including anthocyanins: A review. *Nat. Prod. Commun., 10*, 529–544.

Kumar, S., & Pandey, A. K., (2013). chemistry and biological activities of flavonoids: an overview. *The Scientific World J., 2013*, ID 162750.

Panche, A. N., Diwan, A. D., & Chandra, S. R., (2016). Flavonoids: an overview. *J. Nutr. Sci., 5*, e47.

CHAPTER 12

Colorimetry as a Tool in Plant Physiology

12.1 INTRODUCTION

Light consists of wavelengths of sensible electromagnetic energy to our human eyes, covering approximately in the 380 nm to 760 nm regions. Diverse constituents of plant organs such as chlorophylls, anthocyanins, and carotenoids absorb light, whereas, other uncolored molecules may be converted to pigments with the capacity to absorb light through changes in their chemical structures. The absorption of light, as property, provides the base for quantitative and qualitative analysis of substances with the capacity to absorb light. Colorimeters and spectrophotometers are instruments that measure the amount of light absorbed by molecules relative to wavelengths; these instruments are very sensitive and convenient for quantitative analysis. In these instruments, a chemical substance absorbs light by measuring the intensity of light as a beam of light passes through a sample solution under study. The quantity of pigments may be calculated from the quantity of light absorbed. This absorbance, which also is known as optical density, A, is related to the ratio of the light intensity (I_o) before the beam of light passes through the cuvette with sample solution to the intensity of light which passes through the sample, I (Eq. (1)):

$$A = log \frac{I_o}{I}$$

(1)

The absorbance increases in conformity with the increase in the concentration of pigments. The absorbance also increases with an increase in the distance in which light has to travel through the cuvette containing the solution of pigments.

Different pigments, even thought are present at similar concentrations in identical test tubes or cuvettes, always give different values of absorbance. Thus, a constant of absorption, *a*, should be known for relating the concentration of pigment with measured absorbance. The constant of absorption is in general referred as absorptivity, but also is known as the coefficient of extinction (e) or the coefficient of absorption. This constant depends on the molecular properties of the pigment, the solvent and to a lower extent, to temperature. Since the absorptivity depends strongly on the incident wavelength in the pigment, the colorimeters or spectrophotometers have a mechanism to separate white light of a luminic source (lamp) in diverse wavelength (colors) and direct specific wavelength through the test tube or cuvette which contains the sample.

The absorbance of one solution increases not only by pigment contents, but also by the properties of the tube or cuvette containing the sample and solvent in which the sample is dissolved. Thus, it is necessary to correct the absorbance by the contribution of tube and solvent. For this, a tube or cuvette of identical length and thickness is used, which contains the same solvent but without sample. If Beer-Lambert law is followed (there is a linear relationship between the absorbance and the concentration of a sample), the absorbance (A) is directly proportional to the concentration of pigment (c), the coefficient of absorptivity (a), and length of tube or cuvette with sample (b). This is described in the Eq. (2):

$$A = abc \text{ (Beer-Lambert law)} \qquad (2)$$

This equation has great utility in plant physiology. If the absorbance coincides with absorptivity, as in the case of many molecules of interest to the wavelength specified, it is easy to calculate the concentration of pigment under study. If absorptivity does not coincide not known, this may be calculated from the known concentration of prepared pure pigment.

Finally, the determination of the absorption spectrum, essentially the absorbance at diverse wavelengths, is of great value in identifying many unknown plant components.

12.2 DETERMINATION OF AN ABSORPTION SPECTRUM

12.2.1 OBJECTIVE

To determine the absorption spectrum of plant pigments

12.2.2 MATERIALS

1. One blender with glass vessel;
2. 400 ml of 80% (v/v) acetone;
3. One analytical balance;
4. One Buchner funnel, one Sidearm flask of 250 ml, filter paper and one suction pump;
5. One spatula;
6. One Erlenmeyer flask of 500 ml for the solution of concentrated pigments;
7. One Spectrophotometer and its respective cuvettes. Two graduated pipettes of 1 ml and other of 10 ml.

12.3 METHODOLOGY

Homogenize 5 g (fresh weight) of plant tissue provided by the instructor in 100 ml of 80% (v/v) acetone using a blender. Filter the homogenized tissue material through filter paper kept in Buchner attached to a suction flask and this, to a suction pump. Remove the fragments of tissue adhered to the vessel of the blender and then homogenize with other 100 ml of 80% acetone. Then, extract the pigments still present in the tissue. Filter this homogenized juice in the vessel of blender and filter with other similar portion of 80% acetone. Combine these extracts and mix well. Adjust this volume to 300 ml with 80% acetone. Identify or tag this extract as concentrated solution. Cover the flask to prevent the photo-oxidation of plant pigments.

Be sure the spectrophotometer which has been assigned to you, has been ignited previously in such a form that the indicator of absorbance is stable. Adjust the wavelength to 660 nm. Carefully add 5.0 ml of concentrated solution in one of two tubes or cuvettes and store the rest. Collect 5 ml of acetone at 80% in other tube or cuvette as blank. Adjust the control of absorbance until the indicator gives infinite reading in the scale of absorbance (optical density). Carefully clean walls of tube or cuvette with tissue paper and observe alignment mark of cuvette in the sample holder of spectrophotometer. Thereafter, place the blank cuvette in the sample holder and close the lid. Then, adjust the light control knob until the needle or indicator reads zero absorbance or 100% transmittance. At this stage, a small quantity of light is absorbed by the cuvette and the solvent in blank control, but provides a baseline. This is nullified by adjustment to zero.

Remove the blank from sample holder and close the lid and be assured that the needle or indicator read infinite absorbance. If it is not, the instrument is not balanced, so that it needs to be adjusted to zero with the blank which needs to be repeated one more time. Clean the walls of the cuvette which contains the sample and put in the sample holder. Take the reading of absorbance, and if it is more than 0.4, dilute the solution with one known volume of acetone at 80% till the absorbance is between 0.3 and 0.4 at 660 nm. This guarantee that the absorbance measured in other wavelengths is not beyond of scale. Now it is certain that the instrument is zero with each wavelength fixed, now measure the absorption spectrum of the solution in each 10 nm intervals between 380 and 700 nm. Report graphically absorbance as a function of wavelength (Figure 12.1).

12.4 REPORT

12.4.1 QUESTIONS

1. Which instrument is used for measuring the absorbance of the pigment?
2. Write about the variability observed in the absorption spectrum of different species.

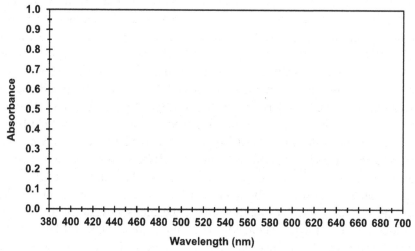

FIGURE 12.1 Absorbance spectrum of isolated pigment from chloroplasts.

KEYWORDS

- **absorptivity**
- **anthocyanins**
- **carotenoids**
- **chlorophylls**
- **colorimetry**
- **spectrophotometer**

CHAPTER 13

Determination of Pigment Contents in Leaf Tissue

13.1 OBJECTIVE

To determine total chlorophyll (*a* and *b*) and carotenoids (xanthophyll and carotene) contents in leaf tissue.

13.2 METHODOLOGY

From the extracts of chlorophyll obtained, prepare one dilution in a form the reading of absorbance at 660 nm is between 0.3 and 0.4 units of optical density. Later, determine the concentration (mg ml^{-1}) of chlorophyll *a* and *b* in the extract by means of using the following equations (Lichtenthaler and Wellburn, 1983):

Concentration of chlorophyll *a*, $Cl_a = 12.21A_{663} - 2.81A_{645}$
Concentration of chlorophyll *b*, $Cl_b = 20.13A_{645} - 5.03A_{663}$

Concentration of carotenoids (xanthophyll and carotene in mg ml^{-1}) determined as follows:

$$Car_{(x+c)} = \frac{1000 * A_{470} - 3.27 * Cl_a - 104 * Cl_b}{229}$$

From the previous equations, determine the quantity (mg g^{-1} fresh weight) of chlorophyll a, chlorophyll b, ratio of chlorophyll a/b, total chlorophyll (a+b), carotenoids (x+c), and the ratio (total chlorophyll)/(carotenoids).

As an example, you may develop a similar graph (Figure 13.1) to illustrate your results for different plant species.

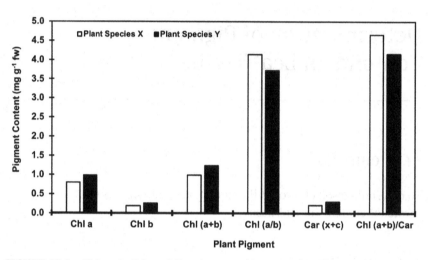

FIGURE 13.1   Chlorophyll (*a* and *b*) and carotenoids content in different plant species.

13.3   REPORT

13.3.1   QUESTIONS

1. Compare your absorption spectrum of pigments isolated from chloroplasts with the values in the textbook for chlorophyll a, chlorophyll b or carotenoids. Explain similarities and differences.
2. Remember that the types of solvent used may influence the absorption spectrum of pigment. Do you consider that the absorption spectrum obtained should be identical with that which is present in the leaf?
3. Do you think that there are differences in the absorption spectrum used in different species of plants?
4. What is the major source of error in this practice?

Include in the report the calculations to determine the contents of chlorophyll (mg g^{-1} fresh weight; mg g^{-1} fw). Besides, generate one graph which represents the contents of each of the pigments. Prepare your results as in other groups of laboratory experiments and interpret:

i. Concentration of chlorophyll *a*: _____

ii. Concentration of chlorophyll *b*: _____

iii. Concentration of total chlorophyll: _____

iv. Ratio of chlorophyll *a/b*: _____

v. Carotenoids (x + c): _____

vi. Total chlorophyll/carotenoids: _____

KEYWORDS

- **absorption spectrum**
- **carotenoids**
- **chlorophyll**
- **chloroplasts**
- **pigments**
- **xanthophylls**

REFERENCES

Lichtenthaler, H. K., & Wellburn, A. R., (1983). Determination of total carotenoids and chlorophylls a and b of leaf extract in different solvents. *Biochem. Soc. Trans., 11*, 591–592.

CHAPTER 14

Water Relations

14.1 INTRODUCTION

Life is intimately related with water, especially in the liquid stage, and its importance for living beings is the result of the physical and chemical properties (Nobel, 1983). Nevertheless, it has been considered as a nutrient for plants in the same way as CO_2 or NO_3^-. Obviously, the quantity of water that is needed for the photosynthesis process is small and only constitutes, approximately 0.01% of the total quantity used by the plant. The cause of this low utilization is owing to the fact that most of the functions which participate are of physical nature. Water is a solvent for many substances such as inorganic salts, sugars, and organic anions and constitutes a medium for all biochemical reactions in plant cells. Water molecules are attracted to surface particles forming layers of hydration which effect the physical and chemical reactions. Water in liquid form permits the diffusion and massive flow of solutes, and for this reason, it is important for transport and distribution of nutrients and metabolites in the plant as a whole. It is important that water present in vacuoles of the cells exert pressure on protoplasm and cell wall, thereby maintaining its turgor of leaves and other plant tissue structures.

14.2 CONCEPTS OF WATER POTENTIAL

The amount of water present in a system is a suitable measure of water status present in plant or soil. The extent of the movement of water, a frequently way for expressing the water status of plants, is called chemical potential, μ. This is a variation of free energy (free energy of Gibbs; $G = H - TS$; a measure of water potential at a point, due to variation, ∂_n of moles of water which enter and come out at this point, being constant other factors (temperature, pressure, etc.). Where:

$$\mu = (\partial G / \partial_n)$$

Water always circulates between two points where thermodynamic potentials are not identical between these two points. The water potential (ψ), used in plant physiology arises of this magnitude. It constitutes the strength of diverse origins (osmotic, capillary, imbibition, turgor, etc.), which contain water in the soil or in different plant tissues. Water potential is the potential energy of water in a system compared to pure water, when both temperature and pressure are kept. Thus, water potential is never positive but has a maximum value of zero, which corrresponds to pure water at atmospheric pressure.

The relation between ψ and μ is described:

$$\Psi = \frac{\mu - \mu^\circ}{V}$$

where: ψ, water potential of sample, units of pressure (Joule/m^3 or Pascal); μ, chemical potential of water in sample, Joule/mol; μ°, chemical potential of pure water, Joule/mol; V: partial molal voulme of water (18×10^{-6} m^3/mol).

The water potential may be expressed as a function of atmospheric pressure in equilibrium with aqueous solution according to the following expression:

$$\Psi = \frac{RT}{V} \ln\left(\frac{e}{e_o} \right)$$

Where R is the universal ideal gas constant (8.32 J/mol/K), T is the absolute temperature (in degrees Kelvin, or K; oC+273.16), and e y e_o, are the atmospheric actual vapor pressure and saturation water vapor pressure of the air, respectively, at the particualr temperature.

In a particular system, total water potential is the algebraic sum of various components:

$$\Psi_w = \Psi_p + \Psi_s + \Psi_m + \Psi_g$$

Where Ψ_p, Ψ_s, Ψ_m and Ψ_g, respectively, components due to the effects of pressure, osmotic, matric and gravitational potential. The component of pressure (Ψ_p) represent the difference in hydrostatic pressure with reference and is positive. The osmotic potential (Ψ_s) results of dissolved solutes, decreases free energy of water, and is always negative. Sometimes is used the term osmotic pressure ($\Pi = -\Psi_s$). The matric potential (Ψ_m) represents the quantity of water bound to the matrix of a plant tissue or soil via hydrogen bonds and is always negative. Ψ_m denotes the affinity of water adsorption to colloidal surfaces colloidal surfaces (solutes or solids). In dry seeds, it can be as low as −2 MPa. In herbaceous plants, this component is approximately −1.0 bar, for which it is ignored. The gravitational potential (Ψ_g) results from the differences in potential energy due to differences in height with respect to the reference level, being positive if it is above the reference and negative when below.

Water potential defines, similarly, the state of water vapor in air, being a function of the relative humidity (HR, %) and temperature of air.

$$\Psi = \frac{RT}{V_W}\ln\frac{HR}{100} = \frac{2.303RT}{V_W}\log\frac{HR}{100} = \Psi = 0.4608 \cdot MPa \cdot K^{-1}(T)\ln\left(\frac{HR}{100}\right)$$

where ψ, is water potential in MPa; R, is the universal ideal gas constant; V is the molar volume of water; T, is temperature in degrees Kelvin; HR, is the relative humidity of air in percentage.

14.3 COMPONENTS OF WATER POTENTIAL (Ψ_w) IN PLANTS

Water potential of plant (Ψ_w), is also used to express the free energy stage of water in cells and plant tissues. Ψ_w consists of three components:

$$\Psi_w = \Psi_p + \Psi_s + \Psi_m$$

Being Ψ_p, Ψ_s, and Ψ_m, the pressure, osmotic, and matric potential, respectively, in one cell or plant tissue.

14.3.1 PRESSURE POTENTIAL (Ψ_p)

When water enters in a cell, increase the volume of vacuoles and exert a pressure, called turgor pressure, on cell walls. This pressure is called wall pressure, acts as an hydrostatic pressure, increase the energy status of water in the cell and represents the pressure potential of cell (Ψ_p). Indeed, Ψ_p acquires positive values only when the vacuole exert a pressure on walls all around.

14.3.2   *OSMOTIC POTENTIAL (Ψ_s)*

This is denoted by the concentration of osmotically active substances in the vacuoles and is identical to the osmotic pressure of cell sap (vant't Hoff equation). In a plant cell, Ψ_s always possesses negative values which vary with cell volume, being near to zero in a fully hydrated cell than in those dehydrated ones.

14.3.3   *MATRIC POTENTIAL (Ψ_m)*

This arises as a consequence of the forces which retains water molecules by capillarity, adsorption, or hydration, occurring fundamentally in cell walls and cytoplasm (matrix).

14.4   MEASURE OF THE WATER STATUS IN PLANTS

Water relation of plants has a physical basis well established, in the sense that the water status may be described both, quantitatively and adequately through water potential. To a large extent, the water potential controls the movement of water in the soil-plant-atmosphere continuum as well as, at the level of cell, tissue and plant organ. An example of the relationship between water status and metabolic porcesses are the adaptation to water stress (drought). It has been demonstrated that some hormones such as abscisic acid (ABA) plays a significant role in these aspects. The most important instruments which are used to evaluate the water status of a plant are: pressure chamber, psychrometers, osmometers, and the pressure probe.

14.5   MEASUREMENT OF XYLEM WATER POTENTIAL WITH THE PRESSURE PUMP

14.5.1   *INTRODUCTION*

During many years, plant physiologists have been limited to study water the water status of plants owing to lack of a simple, rapid, precise, accurate, reliable and portable instrument. Thus, in order to somply with this necessity, Scholander and coworkers in 1965, developed a technique that measures water potential in pressure units. The method consists of applying pressure on an excised and partially sealed leaf till water appears in the extreme end of stem or petiole. At this stage, it is interpreted that the positive pressure applied to the leaf resembles to the pressure which is equal to the negative pressure or water potential maintained inside the stem before cut (Figure 14.1).

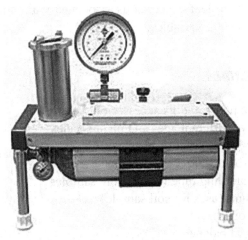

FIGURE 14.1   Determination of leaf water potential through the method of the pressure chamber.

The pressure chamber measures the water potential of the apoplast or cell wall. The matric potential of apoplast (τ) is similar to the water potential (Ψ) of adjacent symplast with the condition that the resistance to water flow between symplast and apoplast is not large enough and the osmotic potential (π) of apoplastic water is approximately zero, which is shown in the following equation:

$$\Psi = P_c - \pi = \tau - \pi$$

Where P_c, the pressure in the pressure chamber. For practical purposes, the osmotic potential of the apoplastic water is lower than 0.5 MPa. Thus,

$$\Psi \approx \tau \approx P_c$$

14.5.2 OBJECTIVE

1. To demonstrate and explain the use of the pressure chamber for the determination of the water potential of plant tissue.
2. To determine and compare the leaf water potential in diverse plant species exposed to conditions of irrigation and water stress.
3. To obtain a diurnal pattern of leaf water potential in diverse plant species.
4. To relate the leaf water status associated with morphological and environmental variables.

14.5.3 MATERIALS

1. Pressure pump and its accessories;
2. Leaves of different ages and plant species
3. A razor-blade cutter or scissors;
4. A razor-blade cutter or scissors;
5. A soil sampling tube for soil core samples
6. Aluminum cases for soil samples
7. Drying oven;
8. Analytical balance

14.6 METHODOLOGY

In order to fulfill the first objective, your instructor explains the mechanism of functioning and the operating instructions of the pressure chamber, with special emphasis on safety precautions of security and he/she will provides the plant materials for the determination of water potential. In the reporting page, register the water potentials obtained for each selected species, leaf age (matured or young), and water status of the medium (irrigated or drought). These determinations satisfy the second objective.

With respect to the third objective, your instructor assigns the plants (replications) of each plant species to study. In order to obtain the diurnal pattern of leaf water potential, determinations of this physiological variable will be carried out at intervals of 2 hours between 06:00 and 18:00 hours of a given date and sampling field plot. Water potential measurements (Ψ)

will be made on terminal twigs of approximately 10 cm long with leaves. Leaf water potential measurements will be taken in three plants per species by using the Scholander pressure chamber at 2 h interval between 06:00 hrs and 18:00 hrs. Environmental variables such as air temperature (°C) and relative humidity (%) could be registered at the same hour sampling time with automated sensors or by using a wet and dry bulb thermometer. Gravimetric soil water content (kg H_2O kg^{-1} soil, dry base) may be determined in soil cores at depths of 0–10, 10–20, 20–30, 30–40 and 40–50 cm using a Veihmeyer type design auger. Each soil sample should be placed in aluminum crucibles. Soil samples will be weighed immediately to obtain the soil wet weight (sww). Later, samples will be placed in a forced air oven at 105°C for 48 h until reaching constant dry weight (sdw). With the help of your instructor, complete the provided tables and build proper figures to illustrate water potential data, environmental variables and relate them. Interpret and discuss widely the water potential as a function of temperature, relative humidity, gravimetric soil water content and whatever other variables which might be important according to your knowledge acquired of plant physiology, bioclimatology, soil science and plant botany. This exercise helps you to comply satisfactorily objective 4 of this laboratory.

14.7 REPORT

TABLE 14.1 Water Potential (MPa) of Xylem in Diverse Plant Species Exposed to Irrigation (Well-Watered) and Water Stress

Plant Species	Plant Number	Sampled Leaf Type		Soil Water Condition	
		Mature	Young	Well Watered	Water Stress

TABLE 14.2   Diurnal Xylem Water Potential (MPa) in (Name of Plant Species) Measured on (Date) at Different Times of the Day

Time of the Day	Replication		
	1	2	3
06:00			
08:00			
10:00			
12:00			
14:00			
16:00			
18:00			

TABLE 14.3   Prevailing Environmental Variables at Different Times of the Day during Water Potential Measurements taken on (Date)

Environmental Variable	Time of the Day						
	06:00	08:00	10:00	12:00	14:00	16:00	18:00
Air Temperature (oC)							
Relative Humidity (%)							
Vapor Pressure Deficit (kPa)							
Dry Bulb Temperature (oC)							
Wet Bulb Temperature (oC)							

TABLE 14.4   Gravimetric Soil Water Content (kg H_2O kg^{-1} Soil Dry Weight) at Different Soil Depths taken on (Date)

Soil Depth (cm)	Soil Water Content (kg H_2O kg^{-1} Soil Dry Weight)
0–10	
10–20	
20–30	
30–40	
40–50	

14.7.1   QUESTIONS

1. What is water potential?
2. What are the different components of water potential?
3. What is the importance to estimate the water potential of xylem tissue?

KEYWORDS

- **hydration**
- **imbibition**
- **relative humidity**
- **turgor**
- **water potential**
- **xylem tissue**

REFERENCES

Nobel, P. S., (1983). Biophysical Plant Physiology and Ecology. W. H. Freeman and Company. New York. 608 pp.

CHAPTER 15

Atomic Absorption Spectroscopy

15.1 INTRODUCTION

Many atomic species have the capacity to emit energy (i.e., radiation) when they are thermally excited or, when they are found in the gaseous state have the ability to absorb electromagnetic radiation. In order to facilitate this absorption, it is required that sample molecules volatilize first and then dissociate in atoms. Thus, to analyze a sample for its atomic constituents, it has to be atomized by means of using a flame (e.g., air-acetylene flame) or by the employment of an electrothermal (graphite tube) atomizer. In the vapor state, free metallic atoms are able to absorb radiation from an optical radiation source such as a hollow cathode lamp, built of the element under study. The amount of absorbed radiation of a particular element (analyte), could be related to the concentration of atoms of that specific element in the sample to be analyzed. This principle requires standards of known analyte concentration to establish the relationship between the measured absorbance at a specific wavelength and the analyte concentration following the Beer's law. Since the source of radiation and the flame are very stable, accurate quantitative determinations of different analytes could be measured with precision in a sample solution.

The utilization of the atomic absorption method consists in the determination of about 70 elements in a sample solution in the range mg L^{-1} to pg L^{-1}. The determination of an element in a given sample of inorganic or organic source, is estimated from a standard curve, statistical procedures, and optimization of technical parameters of the spectrophotometer.

15.2   FUNCTIONS OF ESSENTIAL ELEMENTS

Essential elements have been classified functionally in two groups: those which take part in the structure of an important compound and those which function as activators of enzymes. Carbon, oxygen, and hydrogen are basic examples which involve both functions. Similarly, nitrogen and sulfur, which also are found in many enzymes, are equally important. The magnesium is other example, since it fulfills diverse functions within the plant such as a key central element component in the structure of the chlorophyll molecule and activates many enzymes. The majority of micronutrients are essential owing to the fact that they activate many enzymes. Although, all ions contribute to some extent to lower the osmotic potential and thereby increasing the turgor potential. Potassium and chlorine, both monovalent ions, are also essential elements in plants, which participate as cofactor of many enzymes and it is involved in the oxygen evolution during the photosynthetic reactions, respectively.

The plants respond to an inadequate supply of an essential element by showing symptoms of deficiency characteristics and could have its effect on several parts of a plant. The deficiency symptoms are diverse and can result in stunted growth of roots, stems, leaves, fruits, seeds and, chlorosis and necrosis of leaves. These characteristic symptoms help to determine the role and vital functions of those elements in the plant and growth. These symptoms also help to determine how and when to fertilize the crops. The majority of symptoms appears on buds and young leaves of the plant and could be observed easily. If plants are growing under hydroponic culture conditions, deficiency symptoms in roots could be seen clearly as compared with plants growing under soil conditions either in a greenhouse or in the field. Besides, the symptomos of deficiency differ depending on species, the severity of the problem, stage of growth and the complexities which result in the deficiency of two or more elements. The symptoms of deficiency vary from one element to another. The plant part in which a deficiency symptom occurs depends mainly on two factors: 1) the function or functions of the element in the plant and 2) the mobility of the elements from older plant tissue structures to younger structures. Thus, the symptoms of these elements will appear first in the older parts of the plant. Elements such as calcium, sulphur and iron are relatively immobile. These elements will not be able to be transferred out of the older parts of a plant tissue. Therefore, the symptoms for the deficiencies of these elements will first appear in the younger parts of the plant.

A good example that highlights both factors explained above is the chlorosis (a yellowing of leaf tissue) as a result of the deficiency of magnesium. As magnesium is the central atom in the chlorophyll molecule. This pigment is not synthetized in its absence and only is formed in limited quantities when available at very low concentrations. Besides, the chlorosis of older leaves occurs in the lower part of the plant comes more severe than in younger leaves. This difference illustrates an important principle: the younger parts of the plant has a major capacity to extract mobile elements from older plant tissue structures. Therefore, seeds are especially efficient to extract nutrients, allowing the plant to complete its life cycle.

Some elements are transported through the plant via the phloem from old leaves to younger leaves, and then to plant storage organs. Among these elements, are included nitrogen phosphorus, potassium, magnesium, and chlorine. Other elements such as boron, iron, and calcium are immobile. The mobility of sulfur, zinc, manganese, copper, and molybdenum show an intermediate mobility within the plant. If the element is soluble and also is transported through the phloem, the symptoms of deficiency of this element is perceptible earlier and its deficiency will be more apparent in older leaves, while the symptoms of deficiency of immobile elements such as calcium and iron, symptoms develop first in younger leaves.

15.3 OBJECTIVES

1. To prtactice carefully the procedures of operation of the atomic absorption spectrophotometer.
2. To determine the content of mineral elements in leaf tissue by atomic absorption spectroscopy.
3. To utilize the method of the standard curve.

15.4 METHODOLOGY

Obtain young and mature leaf tissue from the species assigned. Consider three replications of each leaf sample. Leaf samples should be dried to constant weight at 65°C in an oven. Dry samples will be then ground in a mill to pass 1.0 mm mesh sieve. The ground material will be then collected in previously labeled plastic bags with corresponding data for subsequent chemical analysis.

Mineral content will be estimated by incinerating (1.0 g dry weight) ground samples in a muffle furnace at 550°C for five hours. Ashes will be then digested in a solution containing HCl and HNO3 using the wet digestion technique (Díaz-Romeau and Hunter, 1978). Concentrations of Ca (nitrous oxide/acetylene flame), K, Mg, Na, Cu, Fe, Mn, and Zn (air/acetylene flame) will be estimated using an atomic absorption spectrophotometer (AOAC, 1990).

Note: Similarly, at the same time prepare one test tube with reactants using using it as a blank (this did not include sample). One blank for each group is sufficient.

Using the equation $(C_i*V_i = C_f*V_f)$ prepare one standard curve to quantify the concentration of each element in the extract solution from commercial standards.

For the determination of micronutrients (Cu, Fe, Mn and Zn), prepare standards of 0, 0.04, 0.12, 0.2, 0.4, 0.6, 1.0, and 2.0 ppm and measure extract solutions (samples) without dilution.

For the determination of macronutrient (Ca, K, Mg, and Na) content, prepare the standards with concentrations of 0.05, 0.15, 0.45, 0.75, 1.5, and 3.0 ppm and measure the sample using a dilution of 1:200.

After proper calculations, contents of macro (Ca, K, Mg, and Na) and micronutrients (Cu, Fe, Mn and Zn) should be reported in mg Element g-1 dry weight and μg Element g-1 dry weight, respectively.

15.5   REPORT

1. Complete the table with results found.
2. Discuss the differences in nutrient content between mature and young leaves.
3. What factors govern the absorption of elements?
4. Show the possible sources of error in this laboratory exercise.

TABLE 15.1 Leaf nutrient content in mature and young leaves

Plant Species	Element	Mean leaf nutrient content	
		Mature leaf	Young leaf
	Ca		
	K		
	Mg		
	Na		
	Cu		
	Fe		
	Mn		
	Zn		

KEYWORDS

- **chlorosis**
- **spectrophotometry**
- **spectroscopy**
- **volumetric flask**

REFERENCES

AOAC (1990). Official Methods of Analysis. 13th Ed. Association of Official Agricultural Chemists. Washington, DC. pp. 1045–1052.

Díaz-Romeau, R. A., & Hunter, P., (1978). Metodología para el muestreo de suelos y tejidos de investigación en invernadero. CATIE. Turrialba, Costa Rica. Mimeo. 26 pp.

CHAPTER 16

Mineral Nutrition and Symptoms of Deficiency of Nutrients

16.1 INTRODUCTION

Many elements are necessary for a plant to grow, develop and produce seeds. Among these, carbon and oxygen are absorbed as CO_2 and O_2, while hydrogen and some of the oxygen are acquired through absorption of water through roots. The other elements which should be supplied include nitrogen, phosphorus, potassium, calcium, sulfur, magnesium, chlorine, iron, manganese, zinc, boron, copper, and molybdenum. Sodium is considered as a beneficial element for some halophytes and few other elements, simulate growth and may always be tested which are essential for some other higher plant species.

Plants can grow in nutrient solutions under adequated (or limited) concentration of potassium, phosphorus, nitrogen, sulfur, magnesium, calcium, and iron. The symptoms of deficiencies appear in plants and may be studied. Some agricultural experts are familiar with symptoms shown in plants which help to apply fertilizer efficiently; sometimes plant physiologists give guidance on the possible functions in the plants.

The appearance of the deficiency symptoms also gives a guide of the needed element and its movement from one part of the plant to another. For reasons not yet understood, new leaves grow rapidly which removes elements from older tissue strutctures and it is translocated through the phloem. The first leaves formed may receive a good quantity of a given element for supporting normal growth, while newly formed leaf tissue may show deficiency symptoms if the element does not move from old leaves to younger leaves. Similarly, the deficiency symptoms appear early and are more pronounced in younger leaves when the element is slowly transported, while the older leaves exhibit symptoms early in case the elements move promptly.

16.2 OBJECTIVE

1. To demonstrate the criteria of the essentiality of of mineral elements in higher plants.
2. To demonstrate the symptoms of nutrient deficiency for different nutrients.
3. To relate the growth with the symptoms of nutrient deficiency.

16.3 MATERIALS

1. Corn (or bean) plants of 2 to 3 weeks old of three weeks old, grown in plastic pots filled with vermiculite or perlite and irrigated with distilled water.
2. Stock solutions of mineral salts (see Table 16.1).
3. Graduate cylinder of 100 ml.
4. Graduated pipette of 10 ml.
5. 3 light-proof plastic containers of 1 1.0 gallon each.
6. 2 gallons of nutritive solution deficient in one element known (or unknown) to you for K, P, Ca, N, Mg, S, or Fe.
7. pH meter and calibration buffer solutions of pH 4, 7 and 10.
8. Greenhouse or growth chamber.
9. Plastic pots and perlite (or vermiculite) for fillings pots of 250 ml capacity.
10. 4 tags for pots or nutrient solution identification.
11. 2 flasks of 250 ml for helping irrigating plants.

16.4 METHODOLOGY

First, it is essential to prepare the nutrient solutions in which the plants will grow. It needs one gallon in amber plastic or other container, and various stock solutions which are prepared by the instructor and add distilled water to prepare final solutions. The stock solutions are prepared as shown in Table 16.1. The plants will be eventually watered with more dilute solutions which are prepared as follows. Add the quantities of stock solutions as shown in Table 16.2 in one amber plastic container of one gallon which has been half-filled with distilled water. Then, fill the one gallon conmtainer with distilled water and mixed well. For conducting this experiment, you need to be sure that one gallon contains 4 liters, although one gallon contains 3.785 liters (Note: Never use your pipettes with stock solution so that this may contaminate solution with extraneous ions. So that, add in one beaker or graduated cylinder a little more of the quantity required, then measure the exact quantity with a pipette and transfer it to the dilute solution so that the final solution needed is prepared. Discard any remains of stock solution left in the beaker or graduated cylinder; never return it to the original stock solution).

TABLE 16.1 Stock Solutions (Ross, 1974)

Stock solution	Compound	Concentration	g/L
A	$Ca(NO_3)_2.4H_2O$	1.0 M	236.1
B	KNO_3	1.0 M	101.1
C	$MgSO_4.7H_2O$	1.0 M	246.4
D	KH_2PO_4	1.0 M	136.1
E	$Ca(H_2PO_4).2H_2O$	0.01 M	2.52
F	K_2SO_4	0.5 M	87.2
G	$CaSO_4.2H_2O$	0.01 M	1.72
H	$Mg(NO_3)_2.6H_2O$	1.0 M	256.4
I	Minor elements. This solution contains 1.81 g $MnCl_2.4H_2O$, 2.86 g H_3BO_3, 0.22 g $ZnSO_4.7H_2O$, 0.08 g $CuSO_4.5H_2O$, 0.09 g $H_2MoO_4.H_2O$ per liter.		
J	Ethylenediaminetetraacetic acid ferric sodium salt ($Na_2FeEDTA$) prepared by by the instructor. Each milliliter of stock solution should contain 5 mg of Fe		

TABLE 16.2 Milliliters of Stock Solution to be Added in a Container of one Gallon (Ross, 1974)

Treatment	A	B	C	D	E	F	G	H	I	J
	ml of Stock Solution									
Complete	20	20	8	4	0	0	0	0	4	4
-K	30	0	8	0	200	0	0	0	4	4
-P	30	0	8	0	0	80	0	0	4	4
-Ca	0	60	8	4	0	0	0	0	4	4
-N	0	0	2	0	200	80	800	0	4	4
-Mg	20	20	0	4	0	40	0	0	4	4
-S	20	20	0	4	0	0	0	2	4	4
-Fe	20	20	8	4	0	0	0	0	4	0
Distilled Water	0	0	0	0	0	0	0	0	0	0

In a team of two students, a deficient solution of a particular element will be assigned. Th team will be responsible for the growth of the plants irrigated with these solutions and for comparing the deficiency symptoms of the plants grown by your classmates. Adjust the pH of your solution between 5 and 6 with HCl 0.1 N or KOH 0.1 N (or NaOH for the solution deficient in K).

Corn and bean plants will be grown in a greenhouse or growth chamber in plastic pots. Four pots, each 250 ml, will be provided to each team of students. To each pot, a wad of cotton will be placed over each drain hole. Then, fill each pot with perlite (or vermiculite). Sow 4 to 5 seed of the crop assigned on each pot and cover them with 2 cm of perlite. Two pots will be watered with distilled water and two will be watered with the nutrient solution (deficient solution) to be evaluated. Allow the pots to drain out.

Put a tag on each pot, write down the team name clearly and deficient element to be tested. Irrigate daily (50 ml) each pot with nutritive solution since beginning. After seedling emergence, irrigate all plants with 50 ml of water or respective solution daily during the first two weeks for their establishment in each pot. As seedlings (plants) become bigger, watering and nutrient solution requirements increase and these may require more than 50 ml of solution daily. The instructor will advice students to give good irrigation until it drains out.

Now return to the laboratory and calculate the concentrations in ppm (milligrams per liter) and millimoles per liter of Ca, Mg, K, N, P, S, Fe, Mn, Zn, Cu, Cl, B, Na, and Mo in both treatments (complete solution and the corresponding known solution to be evaluated). Two or three significant figures are satisfactory. A table, which illustrates the concentration of each nutrient per solution, have to be included in your report of this experiment of six weeks. During this experiment, make daily observations about the appearance of all the plants and try to identify the unknown solution.

16.5 ACTIVITIES

1. Prepare the solution at the beginning and use it until the experiment is finished.
2. Sow the seeds of the crops assigned.
3. Irrigate the pots daily.

4. Take observations daily from each experiment to evaluate plant gorwth and development.
5. Finally, take the dry weight (DW) of stems and roots from each plant of each treatment and obtain average weight.
6. Ilustrate data in graphs and interpret the results obtained.

16.6 REPORT

For this experiment, each team of students have to write a comprehensive report describing the details of the study, register the daily observations regarding the deficiency symptoms shown by the crop assigned and try to identify the unknown deficiency solution used according to the deficiency symptoms appearance. Relate whether the symptoms of deficiency are associated to mobile or immobile elements. Discuss and share your findings with other teams using the same or different crop and contrast your results with proper literature. What are the major conclusions of this study.

KEYWORDS

- **mineral nutrition**
- **extraneous ions**
- **perlite**
- **stock solutions**

REFERENCES

Ross, C. W., (1974). Plant Physiology Laboratory Manual. Wadsworth Publishing Company, Ing., Belmont, CA. pp. 65-67.

CHAPTER 17

Leaf Epicuticular Wax

17.1 INTRODUCTION

Leaves have a waxy coating. Environmental conditions may strongly affect the amount, composition, and morphology of the waxy coverings of leaf surfaces. Wax and cutin function as main constituents of the leaf cuticle covering the leaf epidermal cells. Epicuticular wax improves the visible and near-infrared radiation, reflectance from leaf surface thus decreasing net radiation and cuticular transpiration and appears to contribute in higher plants to drought tolerance.

17.2 OBJECTIVE

To determine the epicuticular wax content in leaves of different tree species.

17.3 METHODOLOGY

The technique mentioned here is followed by Jacoby et al. (1990) in honey mesquite. Take a sample of about 20 leaves from the middle of leaf canopy from each of five trees selected at random. After collection, store each sample separately in an icebox and take it to the laboratory and stored them until further analysis.

Separate the leaves from twigs before wax removal and form a subsample of these separated leaves of approximately 100 cm² by measuring it with a leaf area (LA) meter. determined by a leaf area meter. Sub-samples were gently rinsed in distilled water to remove foreign material, then air dried, and then placed in a 200 ml beaker with 40 ml of analytical grade chloroform (99% pure) previously heated to 45°C. After

30 seconds, the chloroform was poured into preweighed foil pans which were then placed in a well-ventilated laboratory fume hood and evaporated to dryness for 24 hours. Foil pans were then reweighed to quantify the amount of residual wax. Amount of epicuticular wax for a field sample was determined as an average of five replications. Epicuticular wax content (μg cm-2) is reported on a weight per area basis (Jacoby et al., 1990). Analyze data statistically (Figure 17.1).

[Take a sample of about 20 leaves from the middle of leaf canopy from each of 5 trees selected at random]

↓

[Store each sample of the leaves separately in an icebox and take to the laboratory and store them until further analysis]

↓

[Separate the leaves from twigs before wax removal and form a subsample of these separated leaves of approximately 100 cm² by measuring it with a leaf area meter]

↓

[Wash the subsamples with distilled water to remove adhered foreign material and keep them in a 250 ml beaker]

↓

[Add 40 ml of preheated at 45°C analytical chloroform (99% pure)]

↓

[After 30 seconds, pour out the chloroform in preweighed foil pans, which were placed in a well-ventilated fume hood and evaporated to dryness for 24 hours]

↓

[Reweigh the oil pans to quantify the amount of residual wax]

↓

[Epicuticular wax content is reported on a weight per area basis ($\mu g\ cm^{-2}$)]

↓

[Analyze data statistically]

FIGURE 17.1   Flow diagram to determine epicuticular wax content in foliar tissue.

17.4   REPORT

17.4.1   QUESTIONS

1. What is the variation in the content of epicuticular wax content among the tree species in the experiment?
2. Is there any variability in the epicuticular wax content among the different leaves of the tree species?
3. Specify the importance of the presence of epicuticular wax on the leaf surface?

KEYWORDS

- **twigs**
- **cuticular transpiration**
- **cutin**
- **epicuticular wax**
- **infrared radiation**

REFERENCES

Jacoby, P. W., Ansley, R. J., Meadors, C. H., & Huffman, A.X., (1990). Epicuticular wax in honey mesquite: Seasonal accumulation and intraspecific variation. *J. Range Manage.*, 43, 347–350.

CHAPTER 18

Carbon Fixation (Carbon Sequestration)

18.1 INTRODUCTION

The estimation of carbon fixation/carbon concentration in some trees and shrubs showed that there are some tree species with high capability to fix atmospheric carbon dioxide into their biomass (Gonzalez Rodriguez et al., 2015). In their study, selected trees and shrubs with high carbon concentration were *Eugenia caryophyllata* 51.66%, *Litsea glauscensens* 51.34%, *Rhus virens* 50.35%, *Forestiera angustifolia* 49.47%, *Gochantia hypoleuca* 49.86%, *Forestiera angustifolia* 49.47%, *Pinus arizonica* 49.32%, *Cinnamomum verum* 49.34%, *Bumelia celastrina* 49.25%, *Tecoma stans* 48.79%, *Acacia rigidula* 48.23%, *Eryobotrya japonica* 47.98%, and *Rosmarinus officinalis* 47.77%. A small number of these species may be selected for plantation in highly carbon dioxide polluted areas in cities, roadsides, and factory areas with high emission of carbon dioxide. It is suggested that the tree species with high carbon fixation could be planted in carbon polluted areas to decrease the accumulation of carbon in the atmosphere.

18.2 OBJECTIVE

To determine the carbon fixation ability of tree species by analysis of accumulated carbon in the biomass

18.3 METHODOLOGY

Collect mature leaf tissue samples and place them to dry in a newspaper for a week. Leaf samples should be dried to constant weight at 65°C in an oven. Dry samples will be then ground in a mill to pass 1.0 mm mesh sieve. The ground material will be then collected in previously labeled plastic bags with corresponding data for subsequent chemical analysis. A

2.0 mg of the ground leaf tissue sample will be weighed in an AD 6000 Perkin-Elmer balance in a vial of tin, bent perfectly. This will be placed in a CHONS analyzer Perkin Elmer Model 2400 for determining carbon, hydrogen, and nitrogen. Carbon and nitrogen contents will be reported on a percetage (%) dry weight basis (Figure 18.1).

[Collect mature leaf tissue samples and place them to dry in a newspaper for a week. Dry samples will be then ground in a mill to pass 1.0 mm mesh sieve.

↓

[Dry leaf samples for about 72 h at 65°C in an oven and later place them in desiccators]

↓

[Weigh a 2.0 mg of the sample in an AD 6000 Perkin-Elmer balance in a vial of tin, bent perfectly]

↓

[Carry out in 0.020 g of milled and dried leaf tissue carbon and nitrogen contents (% dry mass basis) by using a CHN analyzer]

FIGURE 18.1   Flow diagram to estimate carbon and nitrogen content in leaf tissue samples.

KEYWORDS

- **carbon dioxide**
- **carbon fixation**
- **CHON analyzer**
- *Eugenia caryophyllata*
- **nitrogen**
- *Rosmarinus officinalis*

REFERENCES

Gonzalez Rodriguez, H., Maiti, R. K., Valencia Narvaez, R. I., & Sarkar N. C., (2015). Carbon and nitrogen content in leaf tissue of different plant species, northeastern Mexico. *Intern. J. Biores. Stress Manage., 6,* 113–116.

CHAPTER 19

Chemical Composition of Wood in Woody Plants

19.1 INTRODUCTION

Wood is a hard tissue below the bark of a tree and is of high commercial importance in the wood industry used for the manufacture of furniture and other domestic uses. Wood quality depends on anatomical structure and chemical composition.

Woody species exhibit variability in percentages of NDF, ADF, lignin, cellulose, hemicellulose, and fiber. The species can be selected on the basis of the high values of each component. The species with a high percentage of NDF, ADF, and fiber may be suggested for the manufacturing of strong furniture, doors, and instruments for domestic uses. The timber of these species may act against shrinkage and strong winds. The species selected for high cellulose and hemicellulose may be recommended for the paper manufacture which may be used for the fabrication of strong furnitures and some for paper.

19.2 OBJECTIVE

To determine the chemical composition of the wood of different tree species.

19.3 METHODOLOGY

19.3.1 PREPARATIONS OF SAMPLES AND TECHNIQUES USED FOR ANALYSIS OF CHEMICAL COMPOSITION OF WOOD

Wood samples of desired tree species will be oven-dried at 60°C during 72 h and ground in a Thomas Willey mill (Thomas Scientific Apparatus, Model 3383) using a N°60 (1 mm × 1 mm) mesh. Save the ground samples of wood and store them in labeled plastic containers. Subject the samples in triplicate for chemical analysis. The chemical composition of

0.5 g of dried (at 60°C for 48 h) milled wood samples will be analyzed for sequential neutral detergent fiber (NDF), acid detergent fiber (ADF) using the ANKOM Fiber Analyzer Apparatus (model 200), and lignin following the guidelines and detailed technical procedures recommended by the Analyzer Operator's Manual. This method evaluates the main fiber constituents including cellulose, hemicellulose, lignin, and ash.

According to Van Soest et al. (1991), the procedures for the determination of the cell wall components are as follows: the NDF (%) is the residue remaining after digesting the sample in a detergent solution to the boiling point for 1 h. The main residues are basically hemicellulose, cellulose, and lignin. With respect to ADF (%), samples are digested with H_2SO_4 (1.0 N) and cetyl trimethylammonium bromide during 1.5 h. In this case, the main fiber residues are mainly cellulose and lignin. Regarding to Acid Detergent Lignin (ADL, %), samples residues are placed in a beaker and treated during 3.5 h in a solution of H_2SO_4 (72%, w/w). Thereafter, samples are washed with water (90°C) until a pH of 7.0 is reached. Then dry samples overnight. Furthermore, samples are washed with acetone and dried at room temperature during 30 min and then dry them at 100°C in an oven during 2 h. Ash content will be determined after incinerating the sample at 550°C during 4 h in a muffle furnace.

The percentage of cellulose and hemicellulose are calculated by using the following equations:

% Cellulose = % ADF - % Lignin
% Hemicellulose = % NDF - % ADF

KEYWORDS

- acid detergent fiber
- acid detergent lignin
- cellulose
- hemicellulose
- lignin
- neutral detergent fiber

REFERENCES

Van Soest, P. J., Robertson J. B., & Lewis, B. A., (1991). Methods for dietary, neutral detergent fiber, and nonstarch polysaccharides in relation to animal nutrition. Symposium: carbohydrate methodology, metabolism, and nutritional implications in dairy cattle. *J. Dairy Sci., 74*, 3583–3597.

CHAPTER 20

Litterfall Chemistry

20.1 INTRODUCTION

Litterfall undergoes different processes of degradation through biochemical processes on the soil surface; finally relase different nutrients, such as carbon, nitrogen, etc. The decomposition of organic matter (OM) plays a central role in the functioning and productivity of terrestrial ecosystems since it determines nutrient cycling and nutrient availability to plants. Research on the decomposition of litterfall is helpful for determining the amount of nutrients that flows in a forest ecosystem from the litterfall reservoir to the forest soil and how they influence the soil fertility. López et al. (2014, 2018), Colín Vargas et al. (2018) and Rodríguez Balboa et al. (2019) have carried out studies on this line of research in different forest ecosystems and the contribution of nutrient content and release has been studied during litterfall decomposition. Nevertheless, many other well documented studies on diverse forest ecosystems and environmental conditions address the release of macro- (C, Ca, K, Mg, N, and P) and micro-nutrients (Cu, Fe, Mn and Zn) during litterfall decomposition.

20.2 OBJECTIVE

To determine the constant rate, nutrient content and release during the decomposition of pine litterfall.

20.3 METHODOLOGY

As an approach to study litterfall decomposition, the litter bag method (Bocock and Gilbert, 1957) will be used. A homogeneous and dried mixture of 10.0 g of freshly fallen litter components (leaves, branches, reproductive structures and other litter components) collected from a forest pine plot, will be deposited in each bag of 25 cm x 20 cm of 1.0 mm nylon mesh fabric. This opening allows the access inside the bags to certain invertebrate detritivorous, but minimizes fragmentation losses (Douce and Crossley, 1982). Litter bags will be randomly placed in the forest soil of a pine stand (25 m x 25 m). On a monthly basis, during a 12-month period, 10 litter bags will be removed and taken to the laboratory and cleaning to remove soil and other foreign material. Then, litter material will be dried at 65oC during 72 h in an oven and determine mass loss. A total of 120 bags will be needed for the study. The litter decomposition rate (a k constant) will be determined through the nonlinear (exponential) Olson's (1963) model defined as $Xt/X0=e{-}kt$; where k is the decomposition constant rate (year-1), Xt is the litter mass (g) at a specific time, Xo is the initial litter mass (g), and t is time (years). The same experimental approach can be replicated in other forest ecosystems and compare the decomposition constant rates. Once the weight of litterfall has been registered, it will be ground in a Thomas Willey mill using a mesh sieve No. 60 (1 mm x 1 mm). The milled material will be stored in plastic containers previously labeled for macro- and micro-nutrient analysis by means of atomic absorption spectrophotometry and/or CHN Analyzer, as documented in previous chapters of this book.

KEYWORDS

- **biochemical processes**
- **decomposition**
- **degradation**
- **litterfall**
- **organic matter**

REFERENCES

Bocock, K. L., & Gilbert, O. J. W., (1957). The disappearance of leaf litter under different woodland conditions. *Plant and Soil, 9*, 179–185.

Colín Vargas, C. I., Domínguez Gómez, T. G., González Rodríguez, H., Cantú Silva, I., & Colín, J. G., (2018). Dinámica de nutrientes durante el proceso de degradación de la hojarasca en el Matorral Espinoso Tamaulipeco. *Revista Mexicana de Ciencias Forestales*, *9*, 87–109.

Douce, G. K., & Crossley, Jr., D. A., (1982). The effect of soil fauna on litter mass loss and nutrient dynamics in arctic tundra at Barrow, Alaska. *Ecology*, *63*, 523–537.

López Hernández, J. M., Maiti, R. K., Gomez Meza, M. V., Gonzalez Rodriguez, H., Cantu Silva, I., Ramirez Lozano, R. G., Pando Moreno, M., & Estrada Castillon, A. E., (2014). Litterfall production and nutrient deposition through leaf fallen in three Tamaulipan thornscrub communities, north-eastern Mexico. *Intern. J. Biores. Stress Manage.*, *5*, 168–174.

López-Hernández, J. M., Corral-Rivas, J. J., González-Rodríguez, H., Domínguez-Gómez, T. G., Gómez-Meza, M. V. & Cantú-Silva, I., (2018). Depósito y descomposición de hojarasca de Pinus cooperi C.E. Blanco en El Salto, Durango, México. *Revista Mexicana de Ciencias Forestales*, *9*, 201–222.

Olson, J. S., (1963). Energy storage and the balance of producers and decomposers in ecological systems. *Ecology*, *44*, 322–331.

Rodríguez Balboa, P. C., González Rodríguez, H., Cantú Silva, I., Pando Moreno, M., Marmolejo Monsivais, J. G., Gómez Meza, M. V., & Lazcano Cortez, J., (2019). Modelos de degradación de la hojarasca en bosques de encino, pino-encino y pino. *Revista Mexicana de Ciencias Forestales*, *10*, 41–55.

CHAPTER 21

Study of the Various Aspects of Mineral Nutrition, Accumulation and Distribution of Micro- and Macro-Elements in Woody Plants

EKATERINA SERGEEVNA ZOLOTOVA

Zavaritsky Institute of Geology and Geochemistry, Ural Branch of Russian Academy of Sciences, 15 Akad. Vonsovsky Street, Yekaterinburg, Russia, 620016, E-mail: afalinakate@gmail.com

21.1 INTRODUCTION

Different methods for research of the various aspects of mineral nutrition, accumulation and distribution of micro- and macro elements in woody plants exist. The selected parts of trees are cut down in three parallels. Vinokurova and Lobanova (2011) took samples of New Year needles and last year's needles of the basic woody species of spruce and fir plantations, thin (diameter less that 1 cm) and thick (diameter 1 cm and more) branches from each one third of the crown. Disks from the trees trunk (diameter at the base of 0.1 m) were cut at the height of 1; 1.3; 3 at every 2 meters. The secondary root of each model tree was dug out for chemical analysis. Ash content was defined by individual burning plant sample in the muffle furnace at the temperature no more than 450°C for 6 hours. The quantitative content of Si, Al, and Fe in samples of plant ash and content of N in absolutely dry samples were determined by the photometric method; the content of Mg, K, Ca, Na – by the method of flame photometry (Methods of biochemical research of plants, 1987). Demakov and his colleagues (2012, 2013) determined the element content in the wood ash (13 different tree species growing in a concise flood plain biotope) with atomic absorption spectrometer "AAnalyst 400." The content of the element in the sample was estimated according to the formula:

$$Ce = Cs \text{ '} \times Vs\text{'} \times Ma/Ms \text{ '} \times Mds,$$

where Ce is the content of the element in the dry sample, mg/kg; Cs – concentration of the element in solution, mg/l; Vs is the volume of the solution in which the ash was dissolved (50 ml for Ca, K, Mn, Zn, Fe, and Cu, and 25 ml for Pb, Ni, Cd and Co); Ma – mass of ash, g; Ms – mass of sample, g; Mds is the mass of the dried sample, g.

X-ray fluorescence analysis is widely used for studying mineral element content in the biomass of various woody plants (Habarova et al., 2015; Tyukavina and Kunnikov, 2015). The mass of one sample (needles, leaves, branches et al.) should no less than 20 g to accurately assess the contents. The samples were analyzed with a wave-dispersive spectrometer "Lab Center XRF-1800."

The content of the most important elements (P, K, Ca, Mg, Sr, Ba, Cl, and Si) in woody plants can be determined by inductively coupled plasma mass spectrometry (ICP-MS) (Vaganov et al., 2013). The measurements were carried out in media prepared by dissolving wood samples in concentrated nitric acid and then diluting them with water (Grachev et al., 2013).

The woody plants shoots can be subjected to wet ashing and determine the content of N, P and K. Nitrogen is determined with the photocolorimetric method using the Nessler reagent. Phosphorus was determined with Truogu-Meyeru method. Potassium was determinated by the method of flame photometry. The elements content was calculated in percentage of absolutely dry mass (Buharina and Dvoeglazova, 2010).

21.2 INVESTIGATIONS OF IMPORTANT PHYSIOLOGICAL PROCESSES

21.2.1 PHOTOSYNTHESIS

The existing methods for determining the productivity of leaf photosynthesis in plants can be divided into the following groups:

1. The gasometrical method with all modifications. For determination of photosynthesis, the air or gas mixture passed through the

chamber with the leaf or plant and determine the concentration of CO_2 in the incoming and outgoing air.

2. The determination of photosynthesis by changing the radioactivity of a gas mixture containing $^{14}CO_2$, which is passed through the chamber with a leaf or plant.

3. The determination of the quantity of oxygen released in photosynthesis.

4. The various modifications of the definition of photosynthesis according to the accumulation of assimilates in plant leaves.

Each of these methods is applied to a greater or lesser extent in various kinds of physiological studies of plants.

Currently, the assessment of the photosynthetic productivity of stands is most often carried out using a modified technique based on the light curves of photosynthesis, obtained in the first half of the day, on different vertical levels (Monsi and Saeki, 1953). The CO_2-gas exchange (net-photosynthesis) is measured by infrared gas analyzers from different manufacturers: "Infralyt-4" (VEB Junkalor Dessau, Germany); "GIP-10MB" (Moscow), as well as portable gas analyzers "LI-6200" (Li-Cor, USA).

The amount of chlorophylls a, b and the sum of carotenoids was determined by spectrophotometry with the use of methodological developments (Shlyk, 1971). Needles in the second year of vegetation with a constant weighed 0.2 g were taken in triplicates during the year. In the laboratory, the needle pigment was extracted in acetone. The extracts of the pigments were filtered by a vacuum method. Optical densities of pigment needle extracts were determined using a single-beam automated spectrophotometer "SF-56" (LOMO) at the absorption centers: for chlorophylls a and b – 644 and 662 nm, for carotenoids – 440.5 nm (Titova, 2010, 2013).

Methods based on the measurement and analysis of variable and delayed fluorescence of chlorophyll sparked interest of scientists in photosynthesis studies (Stirbert et al., 2014; Ptushenko et al., 2014; Alieva et al., 2015). The fluorescent characteristics (fluorescence quantum yield F, maxi-mum fluorescence Fm, quantum yield of photosynthesis Y) of leaves can be measured on a portable chlorophyll fluorimeter "MINI-PAM Yeinz Walz. GmbH" (Alieva et al., 2015).

21.2.2 RESPIRATION

The study of plant respiration is carried out by analogous methods and instruments (gas analyzers), that the study of photosynthesis. However, we found studies where respiration was measured as a heat flux using an isothermal microcalorimeter (Malishev and Golovko, 2011).

21.2.3 TRANSPIRATION

The method of fast weighing, or Ivanov's method is the most simple and convenient method of determining the intensity of transpiration (Ivanov et al., 1950). The method involves weighing cut leaves for short periods of time (1–2 minutes). Changes to the mass of the sheet should be considered prior to its wilting. This method is suitable for the study of diurnal dynamics of the intensity of transpiration, influence of physical factors on water evaporation of plant (Kramer and Kozlovsky, 1983).

Relatively new methods for measuring transpiration in tree trunks have been developed. These methods are based on thermal dispersion. Sensors measure the speed of the xylem sap, which can be correlated to the transpiration of a tree (Belikov and Dmitrieva, 1992).

KEYWORDS

- **chlorophylls**
- **gasometrical method**
- **Nessler reagent**
- **net-photosynthesis**
- **photosynthetic productivity**
- **Truogu-Meyeru method**

REFERENCES

Alieva, M. Y., Mammaev, A. T., & Magomedova, M. H. M., (2015). The photosynthetic characteristics of lighting and shady leaves of woody plants in Makhachkala city. *Proceedings of the Samara Scientific Center of the Russian Academy of Sciences, 17*(5), 67–71.

Belikov, P. S., & Dmitrieva, G. A., (1992). *Plant Physiology* (p. 248). Moskow, RUDN.

Buharina, I. L., & Dvoeglazova, T. M., (2010). *Bioecological Features of Herbaceous and Woody Plants in Urban Plantations* (p. 184). Izhevsk, Udmurt University Publishing House.

Demakov, Y. P., & Shvecov, S. M., (2012). The content of ash elements in annual layers of pine trees in lakeside biotopes of the national park "Mari Chodra." *Eco-Potential, 3/4,* 128–136.

Demakov, Y. P., Isaev, A. V., & Shvecov, A. M., (2013). Consumption and removal of ash constituents from wooden plants in the inundated biotope. *Series: Forest, Ecology, Nature Management, 1,* 36–49.

Grachev, A. M., Vaganov, E. A., Leavitt, S. W., Panyushkina, I. P., Chebykin, E. P., Shishov, V. V., Zhuchenko, N. A., Knorre, A. A., Hughes, M. K., & Naurzbaev, M. M., (2013). Methodology for development of a 600-year tree-ring multi-element record for larch from the Taymir peninsula, Russia. *J. Sib. Federal Univ. Biology, 6*(1), 61–72.

Habarova, E. P., Feklistov, P. A., & Kosheleva, A. E., (2015). Contents of mineral elements in the dying off needles of Scotch pine on drained areas. *Forestry Bulletin, 2*(19), 15–20.

Ivanov, L. A., Silina, A. A., & Cel'niker, Y. L., (1950). About a rapid weighing method for determining transpiration *in vivo. Botanicheskii Zhurnal, 35*(2), 171–185.

Kramer, P. D., & Kozlovsky, T. T., (1983). *Physiology of Woody Plants* (p. 462). Moscow: Publishing House of Forest Industry.

Malishev, R. V., & Golovko, T. K., (2011). The respiration and energy balance in woody plant buds after breaking. *Bulletin of the Institute of Biology of the Komi Scientific Center of the Ural Branch of the RAS, 7/8,* 25–28.

Monsi, M., & Saeki, T., (1953). Uber den Lichtfactor in den Pflanzen ge sells chaften und seine Bedeutung fur dieStoff production. *Jap. J. Bot., 14*(1), 22–52.

Ptushenko, V. V., Ptushenko, O. S., & Tikhonov, A. N., (2014). Fluorescence induction, chlorophyll content and color characteristics of leaves as indicators of aging of the photosynthetic apparatus of woody plants. *Biochemistry, 79*(3), 338–352.

Shlyk, A. A., (1971). Determination of chlorophylls and carotenoids in green leaves extracts. *Biochemical Methods in Plant Physiology* (pp. 154–170). Moscow, Nauka.

Stirbert, A., Riznichenko, G. Y., Rubin, A. B., & Gowinjee, (2014). Modeling of fluorescence kinetics of chlorophyll a: Connection with photosynthesis. *Biochemistry, 79*(4), 379–412.

Titova, M. S., (2010). Content of photosynthetic pigments in needles of *Picea abies* and *Picea koraiensis. Vestnik of the Orenburg State University, 12*(118), 9–12.

Titova, M. S., (2013). Features photosynthetic activity needles introduced species Picea A. Dietr in the arboretum mountain taiga station. *Fundamental Research, 11,* 128–132.

Tyukavina, O. N., & Kunnikov, F. A., (2015). The contents of mineral elements in the phytomass of Scots pine and balsam poplar wood in Arhangelsk. *Arctic Environmental Research, 3,* 80–86.

Vaganov, E. A., Grachev, A. M., Shishov, V. V., Menyailo, O. V., Knorre, A. A., Panyushkina, I. P., Leavitt, S. W., & Chebykin, E. P., (2013). Elemental composition of tree rings: A new perspective in biogeochemistry. *Doklady Biological Sciences, 453*(1), 375–379.

Vinokurova, R. I., & Lobanova, O. V., (2011). Specificity of distribution of macrocells in parts of wood plants of spruce-fir forests in the Republic of Mari El. *Series: Forest Ecology Nature Management, 2,* 76–83.

CHAPTER 22

Allied Microtechniques

22.1 LEAF AREA (LA) AND SPECIFIC LEAF AREA

Matured leaves will be sampled preferably from 10 tree species. Five individuals (replications) from each species and 10 leaves from each individual species, thereby sampling a total of 500 leaves. The leaf area (cm2) can be quantified using a leaf area meter (LI-COR, model LI-3100). Leaf dry mass (g) was taken after drying leaves in an oven at 60oC for 72 hours. Specific leaf area (cm2 g-1) can be calculated as the ratio of leaf area to leaf dry mass (Cornelissen et al., 2003). Another leaf trait characteristics such as leaf width, leaf length, petiole length, and leaf fresh weight can be measured too (Figure 22.1).

22.2 DETERMINATION OF CHLOROPHYLL AND CAROTENOIDS

Five samples of matured leaf tissue (1.0 g of fresh weight) of each plant species can be used for the analysis. The chlorophyll a and b and carotenoids can be extracted in 80% (v/v) aqueous acetone and vacuum filtered through a Whatman No.1 filter paper. Pigment measurements can be determined spectrophotometrically using a UV/VIS Spectrophotometer. Absorbance of chlorophyll a and b, and carotenoids extracts can be determined at wavelengths of 663, 645 and 470 nm, respectively. Chlorophylls and carotenoids contents (mg/g fresh weight) can be calculated according to the equations given by Lichtenthaler and Wellburn (1983). Results are reported on a fresh weight basis (mg of plant pigment per g fresh weight). Total chlorophyll ($a+b$) can be calculated by adding chlorophyll a and chlorophyll b. (Figure 22.2).

22.3 CHARACTERIZATION OF LEAF SURFACE ANATOMY

Leaves contribute greatly to the productivity of plants in a forest ecosystem through photosynthesis, gas exchange (CO_2, O_2), and transpiration. Stomatas play an important role in these vital functions through stomatal pores. The types and size of stomatas vary greatly among species, used in the taxonomic determination of the species along with leaf anatomical features. Leaves among different plant species vary widely in shape, size, margin, and dentation. The epidermis consists of the upper epidermis called adaxial surface and lower epidermis called abaxial surface; it aids in the regulation of gas exchange via stomata. The epidermis is one layer thick, but may possess more layers to prevent transpiration. The epidermis helps in the regulation of gas exchange. It contains stomata, which are openings through which the exchange of gases takes place. Two guard cells surround each stomata regulating its opening and closing. Guard cells are the only epidermal cells to contain chloroplasts.

[Collect ten mature leaves preferably from ten tree species
from each of five individuals (replications) per tree species]

↓

[Collect the leaves in a ziplock plastic bag type and transferred them in a cooler
with ice to the laboratory]

↓

[Take measurements of leaf fresh weight using an analytical balance]

↓

[Measure leaf width, length and petiole length]

↓

[Determine leaf area by means of using a leaf area meter]

↓

[Dry the leaves for three days in an oven at 60°C]

↓

[Determine the dry weight of the leaves and estimate the specific leaf area as the
ratio of leaf area to leaf dry weight]

FIGURE 22.1   Flow diagram for determination of some leaf traits in different tree species.

[Collect about five samples of leaf tissue (1.0 g fresh
weight of each plant species used plant pigment analysis]

↓

[Extract chlorophyll *a* and *b* and carotenoids in 80% (v/v) aqueous
acetone and vacuum filtered through a Whatman No.1 filter paper]

↓

[Pigment measurements can be determined spectrophotometrically using a UV/
VIS Spectrophotometer]

↓

[Determine the absorbance of chlorophyll *a*, and chlorophyll *b*, and carotenoids
extracts at wavelengths of 663, 645 and 470 nm respectively]

↓

Chlorophyll and carotenoids contents (mg/g fresh weight) can be calculated
according to the equations given by Lichtenthaler and Wellburn (1983)

FIGURE 22.2   Flow diagram for determination of chlorophyll and carotenoids.

Leaf anatomical traits play an important role in the taxonomic delimitation of species and its relation to adaptation to the environment such as arid or semiarid conditions. The compact or loose palisade layers also vary. The density of trichomes on leaf surface has been related to the tolerance to insects and drought resistance in crops like sorghum, cotton, etc.

- **Technique 1:** Small pieces of thermocol are slowly dissolved in a small amount of xylene in the Petri dish to bring to a consistency of honey. Then, the solution is applied with the help of little finger on both the upper and lower surface of leaves of each species in the region in between the midrib and margin. Then, they are left on the table for drying. Once dried, a piece of transparent tape is applied and pressed on the region with a finger. Finally, the tape is taken out with a leaf impression and pressed on a slide in the same direction. Now it is permanent and ready to observe under the microscope (Figure 22.3).

[Dissolve small pieces of thermocol in small amount of xylene in the
Petri dish to bring to a consistency of honey]

[Apply the solution with the help of little finger on both the upper
and lower surface of leaves of each species in the region in
between the midrib and margin]

[Leave them on the table for drying]

[Apply a piece of transparent tape on this dried leaf and pressed on the
region with a finger]

[Take out the tape with a leaf impression and press on a slide in the same
direction and observe under the microscope]

FIGURE 22.3   Flow diagram for technique 1 for the study of leaf surface anatomy.

- **Technique 2:** In this technique, a small portion of leaf lamina of
  each species is put in a mixture of 10% chromic acid: 10% nitric
  acid in a test tube and the mouth of the test tube is plugged with
  cotton. Thereby, several test tubes containing several species are
  kept boiling in a water bath in a beaker for 15 minutes. Thereby,
  leaf lamina is clear. Then a portion of the leaf lamina is kept on
  a slide and covered with cover slip with few drops of glycerin
  and stained with a stain and observed under a microscope. The
  anatomical structures observed for the species are presented in the
  following section (Figure 22.4).

By using, any of the techniques takes the impression of leaf surface
(both upper and lower surface) of few species to study the anatomical variability of the species.

For quantification of the stomatal counts and size; stomata may be
counted per mm^2 of leaf surface. With the help of an ocular micrometer,

one can measure the length and breadth of 50 stomatas of each species and count the number of stomata per unit area of leaf surface.

[Place a small portion of leaf lamina of each species in a mixture of 10% chromic acid:10% nitric acid in a test tube]

↓

[Plug the mouth of the test tube with cotton]

↓

[Likewise keep the several test tubes containing several species boiling in a water bath in a beaker for 15 minutes till the lamina is clear]

↓

[Keep a portion of the leaf lamina on a slide and cover with cover slip with few drops of glycerine, stain and observe under the microscope]

FIGURE 22.4   Flow diagram for technique 2 for the study of leaf surface anatomy.

22.4 PETIOLE ANATOMY

Petiole plays a vital role in the transport of nutrients and water to the leaves and provides support to the leaf lamina as well as protruding leaves to solar radiation for photosynthesis. Leaf anatomical traits have been used in the taxonomic delimitation (and adaptation of the species to environments). A number of research scientists have recognized and reported unmistakably the taxonomic importance of epidermal characteristics such as the shape and size of the epidermal cells.

Using a sharp razor blade, section the leaf petiole and midrib samples. Then the thin sections obtained are to be kept in water before transferring onto a glass slide where a few drops of 99% ethyl alcohol was added for tissue hardening and then 2 drops of safranin solutions. Wash off the excess stain with water and add a drop of glycerin. Cover the slides with cover slips and then ring them with nail lacquer to prevent dehydration. Observe the slides with safranin microscope and take the photographs with a digital camera (Figure 22.5).

[Take out the thin sections of leaf petiole and midrib by
using a sharp razor blade]

↓

[Place the thin sections obtained in water before transferring onto a glass slide]

↓

[Add a few drops of 99% ethyl alcohol for tissue hardening
and then add 2 drops of Safranin solution]

↓

[Wash off the excess stain with water and add a drop of glycerine]

↓

[Cover the slides with cover slips and then ring them with nail
lacquer to prevent dehydration]

↓

[Observe the slides with a microscope and take the
photographs with a digital camera]

FIGURE 22.5   Flow diagram to study petiole anatomy.

22.5   VENATION PATTERN, VENATION DENSITY

Leaf venation in a plant represents a typical architectural system of vascular bundle traversing through leaf lamina starting from a petiole. It serves two important functions, offering mechanical strength and transport of water, nutrients, and assimilates, besides hormones. Unlike the three-dimensional stem and root systems of plants, leaf venation can be considered as a two-dimensional ramifying structure. Foliar venation pattern of angiosperms starts with one primary vein, or, more than one primary vein entering the leaf from the petiole and the secondary veins branching off the primary vein(s). Primary and secondary veins are termed the major vein class and represent lower-order veins. The vein diameter at the point of origin of the vein represents the basic criterion in determining the vein order.

Mature leaf samples used for the study can be taken from the upper canopy of different tree species and place them in test tubes containing solutions of H_2O_2: NH_4OH 50%, where each tube contained a particular species to achieve whitening leaf rib and observe in the microscope at 5X. The time required for transparency vary depending on the thickness and leaf contents ranging from 24 hours to a few days. Take photographs with a digital camera fixed in the microscope and count the number of vein islets per unit area mainly on the mid-region of the leaf, between the margin and midrib per unit area at 5X.

22.6 WOOD ANATOMY

Take the wood samples from five trees of woodblock of size 2 × 2 cm and keep them in distilled water for several hours or overnight to soften them for cutting sections with the help of wood microtome at 15 to 2 μm thickness. Take transverse, radial, and tangential sections.

Stain the sections with safranin rapidly for 2 minutes and then wash in distilled water. Then dehydrate the sections with a series of alcohol at 30%, 50%, 70%, 90%, and 100%, making two changes of 5 minutes for each treatment. Finally locate the sections on glass slides, covered with cover slips and fix with euparol to make permanent slides. These sections mounted with euparol are photographed by a digital camera fixed on to the microscope.

Depending on the xylem structure, the sections are in general transversal, radial, and tangential. Transversal section can be cut perpendicular to the length of the trunk. In this plane, it is observed the growth rings and its characteristics, breadth of rings, percentage early and latewood and type of transition among them. Rays can be largely observed as lines which cross the growth ring in the right angle. Other microscopic elements are the type of pores, grouping, and arrangement of pores, size of pores, size of rays, type of parenchyma, texture, and type of transition among soft and heartwood, radial section, perpendicular to the rings. These sections were stained with 2% Safranin-O in water (Jensen, 1962) and mounted with Canada balsam and photographed by a digital camera fixed onto the microscope.

22.7 POLLEN VIABILITY

The phenology and general morphology of pollen of tree species can be studied in general by squashing the anthers and staining the pollens with

1% safranin in water and observe under 40x. For this study, collect the flowers from 11 A.M to 12 A.M for examining the pollen viability. Two techniques are adopted: 1) staining with 1% safranin and 3% iodine in KI, the latter is found to be better than the first one. Separate the fresh anthers, squash, and stain with 3% iodine in potassium iodide for 10 minutes and then count the number of pollen grains stained as viable at 40x under the light microscope.

KEYWORDS

- **abaxial surface**
- **carotenoids**
- **chlorophyll**
- **epidermis**
- **safranin**
- **stomatas**

REFERENCES

Jensen, W., (1962). Botanical Histochemistry: Principles and Practice. W. H. Freeman. San Francisco, CA, USA. 408 pp.

Lichtenthaler, H. K., & Wellburn, A. R., (1983). Determination of total carotenoids and chlorophylls a and b of leaf extract in different solvents. *Biochem. Soc. Trans., 11*, 591–592.

PART III
TECHNOLOGICAL DEVELOPMENTS

Advances in Technology Development in Experimental Ecophysiology and Biochemistry

RATIKANTA MAITI,[1] SAMEENA BEGUM,[2]
HUMBERTO GONZALEZ RODRIGUEZ,[3] and CH. ARUNA KUMARI[4]

[1]*Botanist and Crop Physiologist; Visiting Research Scientist,*
Universidad Autonoma de Nuevo Leon,
Facultad de Ciencias Forestales, Mexico

[2]*Researcher, Department of Genetics and Plant Breeding,*
College of Agriculture, Rajendranagar, Professor Jayashankar
Telangana State Agricultural University (PJTSAU), India

[3]*Faculty Member, Universidad Autonoma de Nuevo Leon,*
Facultad de Ciencias Forestales, Mexico

[4]*Assistant Professor, Department of Crop Physiology,*
Agricultural College, Jagtial, Professor Jayashankar Telangana
State Agricultural University (PJTSAU), India

ABSTRACT

This chapter discusses in brief technology development in various aspects of ecophysiology and biochemistry of trees such as morphology and anatomy, general ecophysiology, specific techniques, formation of growth rings,, growth, and assessment of productivity, miscellaneous techniques, chemical composition of wood, etc.

23.1   INTRODUCTION

Growth, development, and productivity of trees are highly influenced by ecophysiology and biochemistry of trees. Several research works have been carried out to develop technologies to study these traits and its interactions with environments. Here the research advances on these aspects are discussed in brief.

Potvin et al. (1990) acquired ecophysiological response curves from their research trials with replicated measures and performed the statistical analysis. They mentioned that the environmental factors reaction for example temperature, irradiance, water potential, or the concentrations of CO_2, O_2, and inorganic nutrients to physiological or biochemical traits were frequently investigated by physiological ecologists. The data for such a response curve typically are collected by sequential sampling of the same plant or animal, and their analysis should explicitly permit for this statistical design with replicated measures. Normally, in ecology these response curves statistical analysis has either been ignored or improperly assembled.

In an attempt to motivate rigorous analysis of response data, the authors address statistical treatment of response curves and demonstrate the correct alternatives available. For studying response curves, various statistical methods were applied *viz.,* parametric comparison of models fitted to the data by nonlinear regression, analysis of variance with repeated measures (ANOVAR), a nonparametric split-plot analysis (NP split-plot) and multivariate analysis of variance with repeated measures (MANOVAR). For these statistical methods, the C_4 grass *Echinochloa crusgalli,* following chilling, the photosynthesis CO_2 requirement is utilized as an example. Among the various designs, the variance-covariance structure rigid assumptions were made by ANOVAR and considered as possibly the most efficient analysis. Within limits, these assumptions can be relaxed and a corrected significance level used. When the ANOVAR assumptions are not accomplished by variance-covariance structure, the MANOVAR or NP split-plot are feasible substitutes.

The use of MANOVAR in physiological ecology if often restricted because of its small sample size and the treatment factor levels occasionally exceeds the sample size. This problem can be avoided by imparting more attention to experimental design.

de Reffye et al. (1998) examined the uses of modeling plant growth, architecture, and its advances in agronomy and forestry. They stated that in the last two decades, along with modeling of plant structure and growth, two major changes along two major lines have undergone. That is, to encourage the architectural development in a constant and similar environments using morphogenetic models, 3-D virtual plants are generated and the incorporation of ecophysiological information in process-based models in which explanation of plant topology and geometry is frequently absent.

Based on the identification that plant structure: (i) influences the external environment of the trees which itself controls their functioning (competition for space, light attenuation, etc.), (ii) combined result of the physiological processes (water and carbon balance, etc.) and the morphogenetic system of the plant and (iii) directly condition the physiological processes within the tree (hydraulic structure, self-shading, distribution of photosynthates, etc.). Now there is a propensity to combine two approaches that is to associate plant architecture and functioning.

In agronomy and forestry, these models can be employed in different ways: to study the competitive interactions between various plants in the same stand, to observe the water transport and photosynthates distribution within the plant, to analyze the interception of light through the canopy, to examine the local and global, instant, and late effects of the biophysical environment on plant morphogenesis and yield and to standardize remote sensing techniques and to envisage large landscapes.

Yan and Baoguo (2001) considered progress in virtual plant research which involves computer simulations of the growth, development, and plant exploitation in a three-dimensional space. They mention that over the past 20 years, noteworthy progress has been attained in virtual plant modeling pertaining to the rapid advances in information technology and in agronomy, forestry, ecology, and remote sensing areas. These research technologies have wide applications.

An effort to present the importance, method of modeling and main improvements in virtual plant research and applications is made in this review. They also discuss the challenges associated with virtual plant modeling in agronomy applications. These comprise the plant, environment, and root system modeling interaction system. They also mention applications of virtual plants in agronomy in the areas of undertaking virtual experiments to accurately quantify the utilization of soil water and nutrients, to design crop ideotype on computers and to improve crop planting.

23.2 MORPHOLOGY AND ANATOMY

23.2.1 BRANCHING PATTERN

Küppers (1989) reviewed on woody plants above ground architectural patterns of ecological importance. They stated that the growth of woody plants can contribute to various growth forms for instance shrubs and trees through their various branching patterns and their recurring expression while, similar crown shapes also contributes to this. The study revealed that carbon gain, an increase of biomass and its architectural arrangement integration concept is crucial in measuring the cost-benefit association of crown formation and structure, particularly in conditions where crowns compete for light and space.

Maiti et al. (2016) examined the woody plants for their co-existence in a forest ecosystem on the basis of their adaptive morpho-physiological traits. From their results on different morphological, anatomical, and ecophysiological traits they set forth a few hypothetical models for woody plant co-existence and acclimatization in a Tamaulipan thornscrub, Northeastern Mexico.

In the Tamaulipan Thornscrub, Northeast, Mexico for about 30 woody trees and shrubs Maiti et al. (2015) described the possibilities of branching pattern and branching density. They mention that from viewpoints of the ecological perspectives, attention as directed in the modeling of the morphological structure of the plants for developing the model of the functional processes of plants. The branching pattern serves as a solar panel for capturing solar radiation for the production of biomass and timber. They undertook a study to determine the density of branching and types of branching of 30 tree species (trees and shrubs) of the Tamaulipan thornscrub, northeastern Mexico. The study exibited a considerable variability in terms of density and branching patterns. Monopodial, pseudomonopodial, and sympodial forms of branching patterns were noticed. By using animation photography in the field, the branching density has been classified into three types i.e., high, medium, and low density. Large variations are observed in height, biomass, basal trunk and the angle of the primary and secondary branches. With respect to branching density, they categorized three types: high density (15 species), followed by low density (9), and medium density (5 species). The architecture of the tree is the product of the activity of the apical and axial meristems. This model is a strategy for occupying the space and capture of solar radiation.

23.2.2 LEAVES

Ackerly et al. (2002) on the basis of 22 chaparral shrub species dispersal on small-scale gradients of aspects and altitudes, considered the leaf size and specific leaf area (SLA) variations. For the species around the north- and south-facing slopes, the microclimate affinities are measured by evaluating potential incident solar radiation (insolation) from a geographic information system. For the species located in plots along the gradient plots on the basis of their mean trait values at the community level, reduction in leaf size and SLA with rising insolation was observed. However, the leaf size and SLA did not show considerable association across species. This indicated that along this environmental gradient these two traits are decoupled and not related with the same attributes of performance. SLA was extensively lower in evergreen against deciduous species and along the insolation gradient, SLA showed a negative association with the distribution of species. For individual species with respect to insolation distribution, leaf size showed a negative but non-significant trend. With increasing insolation, the variance in leaf size increased at the community level. On south-facing slopes, leaf size exhibited a greater variation for individual species, whereas on north-facing slopes species with small leaves were missing. These results suggest that analyses of plant functional traits along environmental gradients based on community-levels, averages may not reflect the main features of differences in traits and allocation between the component species.

Wright et al. (2005) undertook global-scale quantification of associations between plant traits that give insight into the evolution of the world's vegetation and is necessary for parameterization of vegetation-climate models. They undertook the compilation of database comprising the core 'leaf economics' traits leaf lifespan, leaf mass per area (LMA), dark respiration, photosynthetic capacity, leaf nitrogen and phosphorus concentrations in addition to leaf potassium, photosynthetic N-use efficiency (PNUE) and leaf N:P ratio data for hundreds to thousands of species. It was reviewed that within groups, variations among them were frequently lower than the range observed while the mean trait values varied among the plant functional types. This could be combined with vegetation- climate model in the future. Both globally and within functional plant type's intercorrelation was observed between the main leaf traits thus forming a 'leaf economics spectrum. However, these associations are very common, but they are not universal, showing significant heterogeneity between associations fitted to individual sites.

By variation in sample size, a huge quantity of heterogeneity can be defined. Leaf K and N:P ratios are found to be loosely connected with this trait spectrum while, PNUE is measured as part of this trait spectrum.

Abrams and Mostoller (1995) studied the comparing successional tree species flourishing in open and undergrowth places during drought for their gas exchange, leaf structure and nitrogen content. In central Pennsylvania, USA, during the severe drought on adjoining open undergrowth sites for successional tree species the early (*Populus grandidentata* Michx. and *Prunus serotina* J.F. Ehrh.), middle (*Fraxinus americana* L. and *Carya tomentosa* Nutt.) and late (*Acer rubrum* L. and *Cornus florida* L.) saplings seasonal ecophysiology, leaf structure and nitrogen were estimated. The outcomes unveil that area-based net photosynthesis (A) and leaf conductance to water vapor diffusion (gwv) were highest in open growing plants and early successional species and showed variation among sites and species in open as well as undergrowth sites. The early successional species sunfleck and sun leaves exhibited lesser reductions in A than other species leaves in response to maximum drought period. During the middle and later growing season compared to the sun leaves the shaded undergrowth leaves of all species were found to be more susceptible to drought and had negative midday A values. Further, during the peak drought period, reduced photosynthetic light response was observed in shaded undergrowth leaves.

Compared to shade leaves the sun leaves showed thicker leaves and higher mass per area (LMA) and nitrogen content. The nitrogen content and concentrations in early and middle successional species are more than the late successional species. In both sunfleck and sun leaves, seasonal A showed positive association with predawn leaf Ψ, g_{wv}, LMA, and N and exhibited a negative association with midday leaf Ψ, vapor pressure deficit (VPD) and internal CO_2. Although a significant amount of plasticity was, found in all species for most gas exchange and leaf structural parameters. Among the open and undergrowth plants, middle successional species exhibited then the greatest degree of phenotypic plasticity. Corcuera et al. (2002) studied the functional groups obtained from the pressure volume curves evaluation in *Quercus* species. They mention that oaks that grow on Mediterranean phytoclimates possess common leaf features (ever green-ness, high leaf dry mass per unit area, LMA). It is signified that they might form a coherent functional group due to this phytoclimatic, morphological, and phonological convergence. Through the free transpiration method after calculating pressure-volume curves, curves (*P-V* curves) to confirm this assumptions, some physiological parameters are ascertained. From contrasting phytoclimates seventeen *Quercus* species, six Mediterranean evergreen species (*Q. agrifolia, Q. chrysolepis, Q. coccifera, Q. ilex* ssp. *ballota, Q. ilex* ssp. *ilex,* and *Q. suber*); seven nemoral deciduous species (*Q. alba, Q. laurifolia, Q. nigra, Q. petraea, Q. robur, Q. rubra,* and *Q. velutina*), and four nemoro-Mediterranean deciduous species (*Q. cerris, Q. faginea, Q. frainetto,* and *Q. pyrenaica*) were reviewed. From the homogeneous environmental conditions, i.e., with no water restrictions, uniform light, and nutrient supply two year old seedlings were utilized. The presence of functional homogeneity of the three phytoclimatic groups was verified from the statistical analysis (correlation, mean-value comparisons, and principal component analysis) of *P-V* curves derived leaf features and parameters, which were categorized by their distinct ecophysiological response to water stress. Further, to avoid an excessive loss of cell water, the Mediterranean oak species acquired mechanisms like high cell-wall rigidity. In contrast, the nemoral oaks showed the opposite features. Under water-stress conditions, compared to nemoral oaks the nemoro-Mediterranean oaks works well; however, on relatively dry soils they cannot function as well as the Mediterranean oaks.

Close et al. (2009) studied the ecophysiology of species with distinct leaf morphologies and under a Mediterranean-type climate of hot, dry summers using seedling tube stock ecological restoration. During seedling formation of four co-occurring tree species that vary in leaf morphology the eco physiological influences of plastic tree guards and shade cloth tree guards were examined. It is viewed that the maximum temperature in plastic tree guards was 53.5°C compared to 47.9°C in controls with an average of 6.7°C higher than controls over a summer which partially clarify the results of higher mortality of seedling in plastic tree guards. Relative to control, in summer the light levels were 2-fold lesser and photosynthetic rate was found to be considerably inferior while, midday photochemical efficiency was significantly better in both tree guard treatments. Compared to other treatments, the shade cloth guard trees exhibited a higher photosynthesis and SLA in spring and this raised photosynthesis of seedlings may partly agree with the effect of reduced mortality and improved growth in this treatment. Finally, it is determined that in a Mediterranean type environment shade cloth tree guards gives more favorable microclimate for seedling establishment than plastic tree guards and control treatments. Besides, the environment where low temperature restricts plant growth, these outcomes may have extensive application to the array of restoration settings where seedling tube stock is planted.

Rodriguez et al. (2016) investigated leaf traits biodiversity of woody plant species in Northeastern Mexico. This paper deals with a brief study of research progresses on leaf characteristics, leaf morphology and eco-physiology of trees and shrubs carried out globally and a synthesis of leaf traits in Linares, Northeastern Mexico. Different leaf morphological characteristics *viz.*, leaf dimensions and eco-physiological traits such as leaf area (LA), leaf specific area, leaf dry weight (DW), etc. which based on environmental conditions.

23.2.3 ROOTS

Gale and Grigal (1987) investigated allocations of vertical root of northern tree species in relation to successional status. From 19 published papers for northern tree species measurements of root biomass, number, diameter, and length by soil depth producing a total of 123 vertical root distributions were collected. Further, based on successional status the species were categorized into three tolerance categories. For each excavation a nonlinear function, $Y = 1 - \beta^d$, where Y is the cumulative root fraction from the soil surface to depth d (in cm) was fitted to the data. To test whether the significant variances occur among tolerance classes the regression coefficient β was used as a response variable and considered as an important measure of vertical root distribution. Compared to the late-successional or tolerant species in deeper layers, significantly greater fraction of roots were found in early successional or intolerant species. These differences in vertical root distributions may be because of the inherent genetic ability of early successional species for profound use of a more homogeneous substrate which results from either geologic deposition or nutrient and water redeployment caused after forest disturbance. Owing to their ability to utilize more volumes of soil, early successional species have the capability to adjust to locations restrictive in water and nutrients. Whereas, late-successional or shallow-rooted species are acclimatize to sites where resources are collected near the soil surface as a result of bio cycling and soil development.

23.3   GENERAL ECOPHYSIOLOGY

In the seasonally dry tropics, the deciduous and evergreen woody species ecophysiological characters were analyzed by Eamus and Prior (2001). They state that seasonally dry tropical ecosystems prevail in America, Africa, India, and Australia. Evergreen, deciduous, and semi- and brevi-deciduous trees commonly co-occur and they are biological and cultural conservation spots, regulate regional climate, can support large human populations, and have significant economic value. Research results revealed the mode of responses of these various phenological groups to changes in soil and atmospheric water content. The reason for the existence of evergreen species for a long time while the deciduous species live fast and die young was unveiled by cost-benefit analyses.

To study, recognize, and integrate plant growth and ecophysiology DeJong et al. (2011) applied functional-structural plant models (FSPMs). They mention that these FSPMs models unveil and establish relationships between a plant's structure and processes which direct its growth and development. In recent years, scientists are interested to study the range of topics related to functional-structural plant modeling greatly. Nowadays, the dynamics of growth and development proceeding at the microscopic levels comprising cell division in plant meristems to the macroscopic levels of whole plants and plant communities are enthused by using different FSPM techniques. Broad-spectrum of plant types from algae to trees is investigated. FSPM includes scientists with backgrounds in plant physiology, plant anatomy, plant morphology, mathematics, computer science, cellular biology, ecology, and agronomy forming a highly interdisciplinary research team. Wide-ranging functional-structural models, key processes models, for instance, splitting of resources, plants, and plant environments modeling software, data acquisition and handling systems and utilizations of functional-structural plant models for agronomic purposes, instances are provided by FSPM research team.

Siefert et al. (2015) studied the relative amount of intraspecific trait variation (ITV) in plant communities by conducting a worldwide meta-analysis. They mention that the ITV revealed from the recent findings may deal with the key questions in community ecology. But, in plant communities, the general picture of the relative extent of interspecific trait variation is still missing. In this article, by using data set acquired from 629 communities (plots) and 36 functional traits, the authors conducted a meta-analysis of the relative extent of ITV within and between plant communities globally. On average within communities, 25% ITV of the total trait variation and 32% of the total trait variation among communities is explained. With increasing species richness and spatial extent, the relative amount of ITV decreased however, with plant growth form or climate the ITV did not exhibit any variation. Further, the relative extent of ITV had a propensity to be superior for whole-plant (e.g., plant height) against organ-level traits and for leaf chemical (e.g., leaf N and P concentration) vs. leaf morphological (e.g., LA and thickness) traits. Finally, these results provided practical guidelines for researchers to know when they should involve ITV in trait-based community and ecosystem studies by highlighting the worldwide patterns in the relative significance of ITV in plant communities.

23.4 SPECIFIC TECHNIQUES

23.4.1 *WATER RELATIONS*

Abrams (1990) studied the *Quercus* species of North America to know their adaptations and responses to drought. They mention that because of their capacity either to avoid, or to tolerate, water stress, or both most of the oaks (*Quercus* spp.) in North America are adapted to drought-prone sites. Oaks have water use efficiency owing to their thick leaves and relatively small stomata and they maintain comparatively high predawn water potentials during drought as a result of their deep root systems. But some warm regions species have large stomatal pores which may help in their acclimatization to high temperature by facilitating rapid evaporative cooling. When soil water is abundant the rapid sap movement in large diameter, early-wood vessels and during drought in narrower, late-wood vessels, which are more resis-tant to cavitation slower, but sustained, water movement occurs because of ring-porous xylem anatomy of oaks. At low leaf water potentials and high VPDs, oaks generally have a higher rate of photosynthesis compared to associated species of other genera whereas, *Quercus rubra* normally preva-lent in moderately mesic sites is an exception. *Q. douglasii*, a native of California is drought-deciduous and during drought leaf curling is exhib-ited by some southeastern species while, many oak species, particularly native to arid regions go through modifications in tissue osmotic potential. However, whether such changes are phenological or drought-induced is still to be known. Among species and studies, the published values of bulk modulus of elasticity exhibited large variation. It has been detected that region or changes in predawn water potential or osmotic potential may not be related to increase and decrease during drought. Because of interspecific variation in desiccation avoidance and tolerance during drought, the diurnal leaf water potential may be an insignificant indicator of variations among oak species in the gas exchange rate.

Kubiske and Abrams (1994) undertook the ecophysiological analysis of woody species in contrasting temperate communities during wet and dry years in an intensive sampling regime. During wet (1990) and dry (1991) seasons, by collecting 19 tree species from three different forest communities, they measured leaf gas exchange and tissue-water relations concurrently on the same leaf at midday. The experiment locations were situated on xeric barrens, a misic valley floor, and a wet-mesic floodplain in central Pennsylvania, United States. Xeric, mesic, and wet mesic sites showed a reduction in gravimetric soil moisture by 53, 34, and 27%, respectively, during drought. However, increment in high seasonal mean photosynthetic rates (A) and stomatal conductance of water vapor (gwv) and reduction in midday leaf water potential (ψ) was found in xeric and mesic communities during the wet year, while, wet-mesic community showed low A and gwv and high midday ψ. The mesic and wet-mesic communities exhibited dry year declines in predawn ψ, g_{wv}, and A showing the highest drought effect taking place in the mesic community. Regression analysis revealed that species from each site with high wet-year A and gwv exhibited a tendency to have low midday ψ. But, the mesic community exhibited a reverse trend in the drought year. Inspite of having variances in midday ψ, all three communities had alike midday leaf turgor pressure (ψp) in the wet year due to lower osmotic potential at zero turgor (ψ_π^0) with greater site droughtiness. Instead of high solute content, the lower wet year ψ_π^0 in the xeric community was attributed to low symplast volume. Maybe because of the differences in tissue elasticity species with the lowest ψ_π^0 in the wet year frequently did not include the lowest ψ_π^{100}. Besides, in some species during drought, the osmotic adjustment in ψ_π^{100} may get concealed because of enhanced elasticity but not in ψ_π^0, through dilution of solutes at full hydration. Although the same sampling regime used, there is no relationships existed between gas exchange and osmotic and elastic parameters which were found to be reliably significant between communities or years. This result argues against the universal, direct effect of osmotic and elastic adjustments in the maintenance of photosynthesis during drought.

Therefore, a large number of species should be involved in this study which will provide new insight to the ecophysiology of contrasting forest communities and the widespread consequence of drought on contrasting sites.

Dawson (1996) analyzed the isotopic energy balance and transpiration and the roles of tree size and hydraulic lift to verify the water use by trees and forests. With water use by trees and forests, they measured transpiration (E) rates to estimate the use of soil and groundwater by open-grown *Acer saccharum* Marsh. It was reviewed that the porometer measurements taken from whole canopies constantly miscalculated E by 15–50% while, the Bowen ratio and sap flow methods showed the best results. Different rates of E were displayed by the trees of various sizes. It is supposed that the differential access of large and small trees to groundwater and soil water has led to these differences. It is stated that by locating the water sources based on their hydrogen stable isotopic composition (δD), the transpirational flux was separated between soil water and groundwater. During the entire growing season groundwater, δD was $-79 \pm 5‰$, while soil water δD ranged between -41 and $-16‰$ seasonally (May to September). The large (9–14 m tall) trees showed significantly higher daily transpiration rates than small (3–5 m tall) trees (2.46–6.99 ± 1.02–2.50 versus 0.69–1.80 ± 0.39–0.67 mm day^{-1}). Further, small trees displayed greater variation in E during the growing season than large trees and higher sensitivity to environmental factors like soil water deficits and increased evaporative demand that affect E. Throughout the growing season, large trees and forest stands include trees > 10 m tall which transpires only groundwater. The hydraulic lift (see Dawson, 1993*b*) produces supplemented "pool" of transpirational water in the upper soil layers which influences the high rates of water loss from large trees and older forests. The greater total water flux of large trees is owing to their capacity to use more potential transpirational water during daylight hours than small trees using their hydraulically lifted water reservoir. In contrast, excluding two dry periods when their transpirational water was constituted of between 7 and 17% groundwater the small trees and forest stands composed of younger trees almost entirely exploit soil water. Therefore, compared to small trees and younger forest stands the large trees and older forest stands have a greater effect on the hydrologic balance of groundwater and groundwater discharge from sugar maple trees and forest stands of different sizes (ages) differ significantly. As mixed stands draw water from both soil water and groundwater reservoir they can considerably increase total water discharge on scales from tens to hundreds of hectares. Though mixed stands (small and large trees) may have a higher overall effect on the regional hydrologic balance than old stands.

Gebre et al. (1998) as part of a throughfall displacement experiment at the walker branch watershed near Oak Ridge, Tennessee, during the 1994 growing season considered the changed precipitation effects on leaf osmotic potential at full turgor ($\Psi_{\pi o}$) of some species in an upland oak forest. Overstory chestnut oak (*Quercus prinus* L.), white oak (*Q. alba* L.), red maple (*Acer rubrum* L.); intermediates sugar maple (*A. saccharum* L.) and blackgum (*Nyssa sylvatica* Marsh.); and understory dogwood (*Cornus florida* L.) and red maple species were tested. The treatments were: ambient precipitation; ambient minus 33% of throughfall (dry); and ambient plus 33% of throughfall (wet). High midday leaf water potentials (Ψ_l) in all species in all treatments except in late September, varying from –0.31 to –1.34 MPa for *C. florida*, –0.58 to –1.51 MPa for *A. rubrum*, and –0.78 to –1.86 MPa for *Q. prinus* were observed. Both treatment and species differences in $\Psi_{\pi o}$ were observed, with oak species generally having lower $\Psi_{\pi o}$ than *A. saccharum, A. rubrum, C. florida,* and *N. sylvatica*. The $\Psi_{\pi o}$ of *C. florida* saplings reduced in the dry treatment, and *Q. prinus, Q. alba,* and *A. saccharum* all showed a declining trend of $\Psi_{\pi o}$ in the dry treatment, while $\Psi_{\pi o}$ of *Q. prinus* leaves increased in late August, which corresponded to soil water potential retrieval. *Cornus florida* showed osmotic adjustment with the high adjustment coinciding with the period of lowest soil water potential in June. *Q. prinus*, upheld a lower baseline $\Psi_{\pi o}$ than the other species and it is the only species showing osmotic adjustment. It is determined that in the upland oak forest to change the water relations of some species and to permit the recognition of species capable of osmotic adjustment to a short-term drought during a wet year a 33% decrease of throughfall is sufficient.

Based on some case studies Abrams (1994) made an assessment on genotypic and phenotypic variants as stress adaptations in temperate tree species. It is stated that because of genotypic and phenotypic variation species present on large geographic ranges or a variety of habitats within a limited area are exposed to different environmental conditions. In eastern North America, a series of field and greenhouse experiments involving controlled studies with *Cercis canadensis* L., *Fraxinus pennsylvanica* Marsh., *Acer rubrum* L., *Prunus serotina* Ehrh. and *Quercus rubra* L., in relation to drought stress were performed by the author with students to study these forms of ecophysiological variation in temperate tree species. In this research, they included measurements of gas exchange, tissue water relations, and leaf morphology, and have determined genotypic variation at the biome and individual community levels. During drought, compared to mesic genotypes at incipient wilting, higher net photosynthesis and leaf conductance and lower osmotic and water potentials were exhibited by xeric genotypes. The presence of general coordination among leaf morphology, gas exchange and tissue water relations in xeric genotypes was specified from the presence of leaves with greater thickness, LMA and stomatal density and smaller area than the mesic genotypes. In almost all the study species, leaf phenotypic plasticity to various light environments was obtained signifying a wide array of ecological tolerances. When the interactions of genotypes with the environment were studied, shade plants showed smaller decreases in photosynthesis with decreasing leaf water potential and osmotic adjustment during drought than sun plants. Whereas sun plants displayed significant genotypic variations in leaf structure but with some variables, phenotypic variation surpassed genotype variation. Thus, the phenotypes reacted differently to stress and all phenotypes do not show genotypic variations. On the whole, these studies unveil the importance of genotypic and phenotypic variations as stress adaptations in temperate tree species among both remote and adjacent sites of contrasting environmental conditions compared to 1999, whereas, xeric sites because of their precipitation variability did not exhibit aboveground production variations. In the case of global climate change to produce more frequent occurrences of drought, then the response of mesic sites to prolonged moisture deficiency and the effects of shifting carbon (C) allocation on C storage will raise important questions.

Kolb and Stone (2000) carried out a research in upland, pine-oak forest in northern Arizona to compare leaf gas exchange and water potential among the dominant tree species and major size classes of trees using old-growth Gambel oak (*Quercus gambelii* Nutt.), and sapling pole, and old-growth ponderosa pines (*Pinus ponderosa* var. *scopulorum* Dougl. *ex* Laws.) species. The higher pre-dawn leaf water potential (Ψ_{leaf}) was shown by old-growth oak than old-growth pine indicating the oaks have greater soil water stress avoidance ability. Old-growth oak exhibited higher stomatal conductance (G_w), net photosynthetic rate (P_n), and leaf nitrogen concentration, and lower daytime Ψ_{leaf} than old-growth pine. For pine, at a daytime, Ψ_{leaf} of about −1.9 MPa stomatal closure begins whereas old-growth oak did not show an obvious decrease in G_w at Ψ_{leaf} values higher than −2.5 MPa. In ponderosa pine, P_n, and G_w were highly sensitive to seasonal and diurnal variations in VPD, showing similar sensitivity for sapling, pole, and old-growth trees. On the contrary, P_n, and G_w showed less sensitive to VPD in Gambel oak than in ponderosa pine, exhibiting higher tolerance of oak to atmospheric water stress. Compared with sapling pine, old-growth pine exhibited lower morning and afternoon P_n and G_w, predawn Ψ_{leaf}, daytime Ψ_{leaf}, and soil-to-leaf hydraulic conductance (K_l) and more concentration of foliar nitrogen.

For morning G_w and daytime Ψ_{leaf} pole pine has intermediary values between sapling and old-growth pine comparable to sapling pine for predawn $\Psi_{leaf\ and}$ to old-growth pine for morning and afternoon P_n, afternoon G_w, K_l, and foliar nitrogen concentration. For the pines, low predawn Ψ_{leaf}, daytime Ψ_{leaf}, and K_l were correlated with low P_n and G_w. It is suggested that hydraulic limitations are important in decreasing P_n in old-growth ponderosa pine in northern Arizona and therefore, old-growth Gambel oak exhibits greater avoidance of soil water stress and enhanced tolerance of atmospheric water stress than old-growth ponderosa pine.

Pepin et al. (2002) during wet (1990) and dry years (1990) examined water relations of black spruce trees on a peat land. The benefits given by lower ground-water levels were compromised during drought stress periods. It was seen that seasonal patterns of pre-dawn and mid-day shoot water potentials and stomatal conductance did not have any association with peat water content or to water-table depth. Further, during wet and dry growing seasons no evidence of water stress or osmotic adjustment was noticed in sampled trees. Their soil moisture data revealed that although water-table levels were as low as − 66 cm in 1991, water availability in the root zone was maintained high. During the summer of 1991 a 50% reduction in stomatal conductance appeared in the absence of mid-day water stress when compared with the previous year. They suggest that signals from the bulk of the roots present in the dry peat top layer control the stomatal conductance regulation.

23.5 THE HYDROLOGY OF LEAVES

Sack et al. (2003) analyzed the 'hydrology' of leaves: co-ordination of structure and function in temperate woody species. The hydraulic conductance of the substantially limits whole-plant water transport, but little is known of its association with. The K_{lamina} for sun and shade leaves of six woody temperate species growing in moist soil was measured and its association with the prevailing leaf irradiance and with 22 other leaf traits was studied. K_{lamina} showed variation from 7.40×10^{-5} kg m^{-2} s^{-1} MPa^{-1} for *Acer saccharum* shade leaves to 2.89×10^{-4} kg m^{-2} s^{-1} MPa^{-1} for *Vitis labrusca* sun leaves. Tree sun leaves exhibited 5–67% higher K_{lamina} than shade leaves. K_{lamina} was found to co-ordinate with traits related with high water flux, which included leaf irradiance, petiole hydraulic conductance, guard cell length, and stomatal pore area per unit lamina area. Further leaf lamina thickness, water storage capacity, mesophyll water transfer resistance and in five of the six species with lamina perimeter/area were synchronized with K_{lamina}. However, for six species the leaf traits such as leaf dry mass per area, density, modulus of elasticity, osmotic potential, and cuticular conductance were not inter-related with K_{lamina}. Therefore, K_{lamina} was synchronized with liquid-phase water transport structural and functional traits but independent of other characters involved in drought tolerance and to attributes of sensitivity to pollution.

Wen et al. (2004) based on ecophysiological measurements conducted sensitivity analyses of woody species subjected to air pollution. During the last two decades in the Pearl River Delta of south China air pollution caused a major problem. Natural communities and environments in a wide range of the Delta area were already got damaged because of the emission of air pollutants from industries. Previously, leaf parameters like chlorophyll fluorescence, LA, DW, and LMA were utilized as particular indexes of environmental stress. This study concluded in five seedlings of three woody species, *Ilex rotunda, Ficus microcarpa,* and *Machilus chinensis,* whether the daily variation of chlorophyll fluorescence and other ecophysiological parameters could be used alone or in connection with other measurements for sensitivity indexes to make analysis under air pollution stress and hence to select the suitable tree species for urban afforestation in the Delta area.

After their adjustment under shading conditions, five seedlings from each species were transplanted in pot containers and with a portable fluorometer (OS-30, Opti-sciences, U.S.A) chlorophyll fluorescence measurements were made *in situ*. For LA measurements by area-meter (CI-203, CID, Inc., U.S.A) from each species ten random samples of leaves were selected. Further after drying the leaf samples to a constant weight at 65°C DW was defined and from the ratio of DW/LA, LMA was calculated.

Leaf N content was analyzed according to the Kjeldhal method, and the extraction of pigments was carried out according Lin et al.

Zeppel et al. (2008) investigated a remnant forest during wet and dry years for long term trends of stand transpiration. In a drought year to know the adaptive responses of remnant woodland to drought and to predict the effect of land-use change daily and annual rates of stand transpiration were compared. They adopted two methods for estimations of stand transpiration. First, the E_{sv} method in which the ratio of sap velocity of a few trees measured for several hundred days to the mean sap velocity of many trees measured during brief sampling periods (generally 6–7 trees for 5 or 6 days) were used to scale temporally from the few intensive research periods. Second Penman-Monteith (P-M) equation (called the E_{PM} method) was used. Canopy conductance was predicted by using weather variables and soil moisture which further utilized to determine the daily and annual stand transpiration. The greater E_{PM} values during a drought year and smaller E_{PM} when the rainfall was above average were revealed from the comparisons of daily transpiration estimated with the two methods. Generally, the stand transpiration estimates from two methods were similar and for the two years the E_{PM} method estimated stand transpiration was 379 mm (73%) and 398 mm (37%) whereas, the E_{sv} method gave an estimate of 318 mm (61% of rainfall) in the drought year and 443 mm (42%) in the year having above-average rainfall. The annual stand transpiration from both estimates unveiled that the resilience to an extreme and long-term drought was showed by the remnant forest and it responds rapidly to increased water availability after the drought. Along with this, the annual estimates indicated that groundwater recharge was absent and a larger amount of rainfall was used as transpiration in dry years while the recharge was significantly increased in years with a higher amount of rainfall. Between years, the LA index changes were minimal and variations in stomatal conductance acted as a foremost mechanism for drought adaptation.

23.6 SALINITY TOLERANCE

Cimato et al. (2010) considered salinity tolerance in olive with ecophysiological analysis. Olive (*Olea europaea* L.), an important fruit tree crop species of the Mediterranean basin, is usually exposed to excess soil salinity concentration and in these species mechanisms of response to salinity stress have been studied. This includes essentially (1) the salt-exclusion from the shoot which is also related with a decreased water-mass flow and inherently low relative growth rates 2) the greater use of the mannitol osmoprotectant to partly counter salt-induced decline in leaf osmotic potential. These mechanisms of response cause drastic declines in net assimilation rate leading to less biomass production and thereby, impart adaptive value in olive under salinity stress.

The authors propose that survival, not the growth performance under adverse conditions, confers salt-tolerance in olive, which is subjected to fluctuating soil salinity concentrations during the growing season. During the stress period the olive adopted "low-Na^+ strategy" to deal with excess soil salinity which may be negatively correlated with biomass production. However, on the availability of good- quality water to the roots this strategy allows a fast recovery of plant performance.

Usually with the increase of Ca^{2+} concentration in the root-zone the "low-Na^+ strategy" but the main issue is while the olive grown in calcareous soils, salinity tolerance has been poorly studied in these species. In fact, few data available reveal that high-Ca^{2+} supply increases the capacity of olive plants to recover net photosynthesis and thereby, regain whole-plant growth rate during a period of alleviation from salinity stress, due to the effect of Ca^{2+} in conserving actively growing and old leaves from an enormous accumulation of toxic ions during the stress period.

Finally, they observe that salt-treated olive under natural conditions also get exposed to "excess-light" stress, because the use of sunlight irradiance in photosynthetic processes get drastically declined by stomatal and mesophyll limitation to CO_2 diffusion in the leaf. The role of major biochemical adoptions viz. changes in mannitol, violaxanthin-cycle pigment, and flavonoid concentrations which counteracts the oxidative damage triggered by the rigorous action of salinity stress and high sunlight were discussed. To protect the salt-treated leaves from heat stress-induced oxidative damage in full sunshine to a greater magnitude than leaves growing under partial shading these adjustments have been recommended. In the ecology of *O. europaea*, these mechanisms play an important role in the response systems to reduce salinity stress in this species.

23.7   MODELING OF CARBON DIOXIDE

During the 1995 summer drought, Baldocchi (1997) considered and demonstrated carbon dioxide and water vapor exchange over a temperate broad-leaved forest. He state that during the 1995 growing season forests in the south-eastern United States were exposed to a prolonged dry spell and above-normal temperatures. During this period, throughout a temperate, hardwood forest nearly continuous, eddy covariance measurements of carbon dioxide and water vapor fluctuations were assembled. These data are used to test the environmental factors and accumulated soil moisture deficits effect on the diurnal pattern and magnitude of canopy-scale water vapor and carbon dioxide fluctuations. To verify an integrative leaf-to-canopy scaling model (CANOAK), the field data was used. This model provides the estimates of mass and energy fluxes using micrometeorological and physiological theory. During the spring, when soil moisture was abundant, peak rates of net ecosystem CO_2 exchange (N_F) which occurred around midday exceeded 20 μmol m^{-2} s^{-1}. Rates of N_K reached near optimal when air temperature varied between 22 and 25°C. Earlier in the morning, the accumulation of soil moisture deficits and a co-occurrence of high temperatures induced maximum amount of daytime carbon dioxide uptake. High air temperatures and soil moisture deficits showed correlation were also associated with a drastic decline in the extent of NE. It was observed that on average, the extent of N_E reduced from 20 to 7 μmol m^{-2} s^{-1} owing to an increase in air temperature from 24 to 30°C and the dry soil. When the adequate supply of soil moisture takes place in the forest, the precise estimates of canopy-scale carbon dioxide and water vapor fluxes can be known from the CANAOK model. During the drought and heat spell, a cumulative drought index was necessary for adjusting the proportional constant of the stomatal conductance model to give accurate estimates of canopy CO_2 exchange. Further, to provide advanced estimates of evaporation until midday, the adoption of the drought index helps the CANOAK model. Conversely, the scheme could not provide precise estimates of evaporation during the afternoon.

23.7.1 PHOTOSYNTHESIS

Gunderson and Wullschleger (1994) made a leaf level photosynthetic responses analysis using 39 tree species grown in elevated concentrations of atmospheric CO_2 in which at the growth $[CO_2]$ on an average 44% more photosynthetic rate was observed. The photosynthesis of plants grown at elevated $[CO_2]$ found to be declined on average 21% when compared with the photosynthesis measured at a common ambient $[CO_2]$. However, variability was high. The anatomical and morphological changes in trees that effect leaf- and canopy-level photosynthetic response to CO_2 enrichment accompanied by the proof relating the photosynthetic acclimation in trees with variations at the biochemical level was tested. The evidence in trees for one predominant factor controlling acclimation is not known. However, the variations in sink strength and nutrient strength appear to affect photosynthetic acclimation. Regardless of the mechanisms that influence photosynthetic acclimation, it is uncertain that this response will be complete. In order to predict ecosystem response to a changing environment a novel effort on adjustments to rising $[CO_2]$ at the canopy, stand, and forest scales is essential.

Rascher et al. (2000) evaluated the instant light- response curves of chlorophyll fluorescence parameters acquired with a portable chlorophyll fluorometer on site in the field. The miniature pulse-amplitude modulated photosynthesis yield analyzers were formed which were mainly designed for determining effective quantum yield ($\Delta F/F_m'$) of photosystem II under momentary ambient light conditions in the field. Although this gives important ecophysiological information, it is often required to learn more about the potential intrinsic capacities of leaves by measuring light-response curves. Thus, instruments give light-curve programs, where light intensities are enhanced in short intervals and instant light-response curves are noted within a few minutes. This method is open to criticism because photosynthesis is probably not being in steady state. With the suitable precautions, this technical data specified that the instant light curves can nonetheless provide reliable data about cardinal points of photosynthesis. First, the geometry of the light source of the instrument relative to the quantum sensor has to be measured and quantum sensor readings should be adjusted. Second, under ambient light conditions where photosynthesis is in stable state, the measurements of the light-response curves should be equated with readings of effective quantum yield of photosystem II. This may indicate that either both measurements perfectly concur or offset against each other by a constant value within the critical range of the light curves. In the first case, outcomes of light curves can be taken at face values and in the second case; a simple adjustment can be applied. A huge benefit in ecophysiological field work can be acquired from instant light-response curves with these precautions and careful interpretations.

23.7.2 RESPIRATION

Sprugel (1990) explored constituents of woody-tissue respiration in young *Abies amabilis* (Dougl.) Forbes. At three to five locations on each of 12 30-year-old *A. amabilis* trees woody-tissue respiration was measured on five different dates. Between sampling locations temperature-corrected respiration per unit surface area showed variation 10 to 40-fold on any given date. The growth respiration and sapwood maintenance respiration are the two major constituents of respiration in stem and were of almost equal significance during the growing season. The evidence of significant cambial maintenance respiration was not seen indicating that a stand with high bole surface area would not spontaneously have high respiration. Respiration showed a significant association with branch height and in boles of comparable volume and growth rates, the respiration was much lower than in branches. Along with the two components observed in bole respiration, another important component possibly correlated with carbohydrate mobilization and transport or with CO_2 efflux from the transpiration stream may involve in branch respiration.

23.8 FORMATION OF TREE RINGS

Hättenschwiler et al. (1996) studied the effect of elevated CO_2 and increased N deposition on tree ring response in *Picea abies*. In a simulated mountain forest climate three CO_2 concentrations (280, 420 and 560 cm^3 m^{-3}) and three rates of wet N deposition (0, 30 and 90 kg ha^{-1} $year^{-1}$) were applied to four to seven-year-old spruce trees (*Picea abies*) for 3 years. Further, in nutrient-poor natural forest soil, six trees from each of six clones were grown in competition in each of nine $100 \times 70 \times 36$ cm model ecosystems. Afterwards, using X-ray densitometry stem dices were analyzed. It was found that the growth in the radial stem was not influenced by [CO_2] but it increased with increasing rates of N deposition. Wood density was improved by [CO_2], but reduced by N deposition. Wood-starch concentration enhanced and wood nitrogen concentration reduced with increasing [CO_2], but neither was affected by N deposition. Moreover, [CO_2] or N deposition did not cause any effect on lignin concentration in wood. In conclusion, it is proposed that increasing atmospheric [CO_2] will not increase radial stem growth of spruce under natural growth conditions, however; atmospheric N deposition will do and in some regions is probably doing so. Except this effect is compensated by enormous atmospheric N deposition elevated [CO_2] will produce denser wood. In forest ecosystems, the mechanical properties of wood may change owing to the greater wood density under elevated [CO_2] whereas, higher ratios of C/N and lignin/N in wood grown at elevated [CO_2] may affect nutrient cycles.

Brooks et al. (1998) considered indications from tree-ring widths and carbon isotope analyses to study the responses of boreal conifers to climate fluctuations. According to them, partial distribution and species composition of the boreal forest are expected to change under predicted climate change scenarios. The research findings reveal that water deficit control the southern boundary of the central Canadian boreal forest and temperature limitations control the northern boundary. As part of the Boreal Ecosystem, the annual climatic parameters in the northern and southern boreal forest are compared with annual variation in tree-ring widths and carbon isotope ratios (δ ^{13}C) of tree-ring cellulose. Contradictory to expectations, at the northern and southern sites, climate correlations with ring widths were alike in black spruce (*Picea mariana* (Mill.). Cooler and wetter environments supported annual growth. For jack pine (*Pinus banksiana* Lamb.) enhanced temperature and spring precipitation favored annual growth at both sites. In the north, annual growth showed a negative association with winter precipitation.

The current distribution theories were supported by the δ^{13}C – climate correlations in *Pinus banksiana*. Winter and growing season precipitation affected annual δ^{13}C variations in the north while, in the south, potential evapotranspiration produced significant annual δ^{13}C variation. The conception that the cold soil temperatures restrict the northern extent and moisture restricts the southern range of *Pinus banksiana* was supported from their data. Conversely, colder, and wetter conditions supported the growth of *Picea americana* all over its range. These interpretations strengthen the concept that species respond individually to climate change, not as a cohesive biome.

Abrams et al. (1998) studied tree-ring responses to drought throughout species and contrasting sites in the Ridge and Valley of Central Pennsylvania. To ascertain differential drought impacts on xeric ridge, dry-mesic barrens, mesic valley, and wet-mesic riparian sites they quantified annual tree-ring variation between 1984 and 1995 in 10 hardwood species in central Pennsylvania (USA). Each tree species grew on two or more sites which were subjected to four droughts of moderate to severe intensity that occurred during the study period. It was found that the trees on the mesic valley and dry-mesic barrens sites had experienced less growth reduction while, trees on the xeric ridge and wet-mesic riparian sites were usually influenced with a decrease in the ring width index. Annual variation in growth showed a negative correlation with summer temperature, particularly found for species on the riparian site. During one or more of the droughts out of twenty-three species-by-site combinations six species showed above-average growth, whereas 8 of 23 species-by-site combinations had below-average growth the year following drought. In *Acer rubrum* in the valley, *Prunus serotina* in the barrens and riparian sites and *Quercus rubra* and *Acer rubrum* on the ridge the species with the largest decline in the ring width index during drought exhibited variations among the sites. The data got by comparing tree-ring data of this study with leaf gas exchange data from other studies provided several inconsistencies in drought responses. Thus, information from both methods may be required in precisely measuring species drought tolerance rating.

Carrer and Urbinati (2004) verified the consistency of climate-growth responses in tree-ring series from *Larix decidua* and *Pinus cembra* trees of four age classes. In *P. cembra* with age, tree-ring statistics (mean sensitivity, standard deviation, correlation between trees and first principal component) did not differ significantly but in *L. decidua*, they seemed to be related with age classes. Response function analysis has shown that climate gives details for a high amount of variance in tree-ring widths in both species. The older trees showed the higher variance explained by climate, the significance of the models and the percentage of trees with significant responses. Age influence on climate sensitivity is possibly to be non-monotonic. In *L. decidua* according to a twofold pattern, the most important response function variables changed with age *viz.,* increased for trees younger than 200 years and declining or constant for older trees. A comparable pattern was detected in both species for the association between tree height and age. It is presumed that in trees with growth and age an endogenous parameter connected to hydraulic status gradually gets limiting encouraging more stressful conditions and a higher climate sensitivity in older individuals. The climate signal is optimum in older trees and sampling method is non-stratified by age (especially in multi-aged forests) might be biased mean chronologies because of the higher amount of noise present in younger trees were confirmed from the findings of this study. This needs more extensive research as there are significant ecological implications both at small and large geographic scales. If the age effect is not accounted for predictive modeling of forest dynamics and paleo-climate reconstructions might be less robust.

Rozas (2005) examined old-growth woodland in the Cantabrian lowlands, northern Spain to study both time-independent and time-dependent responses of radial growth in pedunculate oak (*Quercus robur* L.) to climate using correlation analysis, bootstrapped response functions and Kalman filter analysis. Three different oak age-classes: young < 120 years, mature 170–225 years old, and old-growth trees 250–470 years were considered for responses to climatic factors. Age-specific responses were shown from time-independent climatic models and between 55.2 to 66.8% of ring-width, experimental variance was elucidated by weather conditions in the period 1940–1998. The mature and old-growth oaks displayed a negative response to winter and summer temperatures and a positive one to summer precipitation, whereas in young oaks radial growth was limited by temperatures in June of the year the ring was formed. Time-dependent models demonstrated that the influence of some climatic variables on tree growth was not constant through time. This variability could be accredited to physiological changes associated with tree aging though it was not associated to changes in other environmental factors. The results revealed the expected probable responses of oaks growing in old-growth to global climatic changes and indicated that any hypothesis of an age-independent climate-growth relationship for oak in the experimented locality is not reliable.

Büntgen et al. (2007) examined the multi-species tree-ring network growth responses to climate in the Western Carpathian Tatra Mountains, Poland, and Slovakia. In this study, the growth of 24 tree-ring width and four maximum latewood density chronologies were analyzed. From four conifer species (*Picea abies* (L.) Karst., *Larix decidua* Mill., *Abies alba* (L.) Karst. and *Pinus mugo* (L.)) between 800 and 1550 a.s.l. 1,183 ring-width and 153 density measurement series were collected to form this network and the network was analyzed to determine growth responses to climate as a function of species, elevation, parameter, frequency, and site ecology. In order to retain annual to multi-decadal scale climate information in the data individual spline detrending was used. Further, from 16 grid-boxes covering the 48–50 N and 19–21 E regions, twentieth-century temperature and precipitation data was used for comparison. The latewood density chronologies exhibited a correlation with the April-September temperatures while twenty ring-width chronologies was significantly correlated ($P < 0.05$) with

June-July temperatures. To explain the growth response to climate site eleva-tion and frequency of growth variations (i.e., inter-annual, and decadal) were found to be significant variables while, climatic effects of the previous year summer did not had any significant effect on ring formation and the response to precipitation was increased with decrease in elevation. For *Larix decidua* association between summer temperatures and annual growth was less than for *Picea abies.* Five dominant eigenvectors showing somewhat contrasting climatic signals were identified from principal component analysis. 42% of the network's variance was described by the first principal component with highest loadings from 11 *Picea abies* ring-width chronologies and one *Pinus mugo* ring-width chronology. The mean of three latewood density chronolo-gies, loaded most strongly on the fourth principal component, was signifi-cantly correlated at 0.69 with April-September temperatures ($P < 0.001$ over the 1901–2002 period in both cases) but the mean of these 12 high-eleva-tion chronologies showed significant correlation with 0.62 with June-July temperatures. These groupings let strong assessments of June-July (1661–2004) and April-September (1709–2004) temperatures. The general rule of the key influence of growing season temperature on high-elevation forest growth was supported from the comparison with reconstructions from the Alps and Central Europe.

Vieira et al. (2009) investigated tree-ring growth and intra-annual density fluctuations responses of *Pinus pinaster* to Mediterranean climate. Dendrochronology usually predicts that climate-growth associations are age-independent once the biological growth trend has been eliminated. Whether the radial-growth response to climate and the intra-annual density fluctuations (IADFs) of *Pinus pinaste* showed or not was tested. In Pinhal de Leiria (Portugal), trees were sampled and divided in two age classes: young (<65 years old) and old (>115 years old). It was found that the response of latewood width to climate was greater in old trees whereas early wood and tree-ring width of young *P. pinaster* trees were more sensitive to climate effect. The growing season of young trees begins earlier so, there occurs delay between young and old trees during these wood cells of young trees assimilate environmental signals. Further, young *P. pinaster* trees produce a greater frequency of IADFs than old trees because they generally have a longer growing season and respond faster to climate conditions. The radial-growth response of *P. pinaster* to climate and the IADFs frequency were age-dependent and maximum IADFs were present in latewood exhibiting positive association with autumn precipitation. The resolution of climatic signals can be improved by using trees with different ages producing a tree-ring chronology for climate studies and finally in what way young and old trees react to climate change can be predicted from these age-dependent responses to climate.

23.9 GROWTH AND ASSESSMENT OF PRODUCTIVITY

Ritchie (1984), published a manual on assessing seedling quality characteristics of planting stock which reflect quality (defined here as performance potential). The seedlings were classified either as "performance" attributes or "material" attributes. By exposing whole seedlings to certain environmental treatments and their responses to some performance attributes, like root-growth potential, cold hardiness, and stress resistance, were measured. Performance attributes tend to need laborious and time-consuming procedures but they are integrators of all or many seedling subsystems and frequently correlate well with seedling performance potential. By using any number of direct or indirect methods, the attributes in question were measured and material attributes, such as dormancy status, water relations, nutrition, and morphology were estimated. Except values fall well outside of some established range, material attributes generally provide less definitive information on seedling quality but they can be more easily and rapidly measured than performance attributes. But most methods were employed to indicate the desirability of carrying out some cultural operations, for instance, irrigation or lifting, instead of measuring seedling quality itself.

Ceulemans and Mousseau (1994) analyzed the elevated atmospheric CO_2 effects on woody plants. They mention that owing to important role of trees in the global carbon balance and their possible carbon sequestration with respect to global climatic changes, the acquaintance with these processes is necessary for understanding the functioning of the entire forest ecosystem which can be modeled and predicted based on the physiological process information. The principal methods and techniques employed to study the probable impacts of elevated CO_2 on woody plants and similarly on the major physiological responses of trees to elevated CO_2 were reviewed by the authors. The available exposure techniques and methods are explained and the percentage changes in biomass, root/shoot ratio, photosynthesis, and LA and water use efficiency under elevated CO_2 were summarized from an overview table with all relevant literature data over the period 1989-93. With a specific emphasis on downward regulation of photosynthesis, the interaction between growth, photosynthesis, and nutrition was considered. They also discussed stimulation or reduction observed in the respiratory processes of woody plants along with the elevated CO_2 effect on conductance, water use efficiency and stomatal density. Variations in plant quality and their significances are observed. Further changes in underground processes under elevated CO_2 are especially strengthened and related to the functioning of the ecosystem.

Newman et al. (2006) assessed above- and below-ground net primary production in a temperate mixed deciduous forest. They mention that a poor understanding of belowground productivity, the short duration of most analyses and a necessity for greater investigation of species- or community-specific variability in productivity studies, limits the current capability to identify and predict variations in forest ecosystem productivity. In both mesic and xeric site community types of the mixed mesophytic forest of southeastern Kentucky, over 3 years aboveground net primary productivity (ANPP) and both belowground NPP (BNPP) and total NPP over 2 years (2000–2001) were measured to verify landscape variability in productivity and its association with soil resource [water and nitrogen (N)] accessibility. Around sites, ANPP exhibited significant correlation with N availability ($R^2 = 0.58$, $P = 0.028$) whereas BNPP was best projected by soil moisture content ($R^2 = 0.72$, $P = 0.008$). Because of these counteracting patterns, total NPP was not associated to either soil resource. Interannual variability in growing season precipitation during the study gave a 50% reduction in mesic site litter production, possibly due to a lag effect following a moderate drought year in 1999. As a result of which ANPP in mesic sites reduced to 27% in 2000.

Canham (1988) reviewed the growth and canopy architecture of shade-tolerant trees: in saplings of *Acer saccharum* Marsh. (sugar maple) and *Fagus grandifola* Ehrh response to aboveground growth canopy gaps patterns, branching, and leaf architecture was investigated to know: (1) the aboveground growth rates responses to change in forest light regimes and (2) the significance of branching and leaf display patterns to the capability of saplings of these two species to respond to variations in forest light regimes created by canopy gaps. For both species, in the low gap light levels produced by small canopy gaps (15–75 m^2) rate of height growth, lateral growth, and the production of new shoots were as much as an order of extent greater than growth rates of saplings underneath closed canopies. But to the additional increase in light gap levels saplings of both species showed little response. An increase in sapling LA index and the effectiveness of leaf display (measured as LA per unit length or surface area of branches) were related with the strong response of maple saplings to low gap light levels. Absence of significant increase in beech LA indices in small gaps and a higher efficiency of leaf display underneath a closed canopy than in small gaps were corresponding to the more uncertain response to low light levels in small gaps and the higher growth rates than maple underneath a closed canopy. Therefore, in both of these species the degree of plasticity in patterns of branching and leaf display found to be related to the extent of the response to small canopy gaps. Further, both species can be regarded as small gap experts owing to their combinations of shade tolerance, growth responses, and canopy architecture which are particularly successful to exploit small canopy gaps. However, the two species vary in their placement on a gradient in the degree to which woody plants respond to canopy disturbances through the production of a new flush of leaves and increased stomatal conductance.

Donovan and Ehleringer (1991) assessed ecophysiological variations among juvenile and reproductive plants of several woody species. For four woody species local to Red Butte Canyon, Utah, USA, such as *Acer negundo, Artemisia tridentata, Chrysothamnus nauseosus*, and *Salix exigua* photosynthetic and water relations features of small juvenile and large reproductive plants during one growing season were studied. At least one of the following characters: water potential, stomatal conductance, photosynthetic rate, or water-use efficiency was showed by reproductive plants of juvenile plants for all species while late in the growing season mortality was seen in the juvenile plants evidently because of lack of water. However, within reproductive plants, such mortality was not observed. The observed variances between juvenile and reproductive classes were in terms of environment, development, and mortality rates.

23.10   EFFECT OF OZONE

Wittig et al. (2009) measured the current and future tropospheric ozone effect on tree biomass, growth, physiology, and biochemistry using: a quantitative meta-analysis. Presently important carbon sinks are the northern hemisphere temperate and boreal forests but, current rising concentrations of tropospheric ozone ($[O_3]$) and $[O_3]$ projected for later this century are causing damage to trees and have the potential to decline the carbon sink of these forests. The current $[O_3]$ and future $[O_3]$ impacts on the biomass, growth, physiology, and biochemistry of trees prevailing in northern hemisphere forests were verified by meta-analysis. When compared with trees grown in charcoal-filtered (CF) controls, representing estimated preindustrial $[O_3]$ the current ambient $[O_3]$ (40 ppb on average) significantly decreased the total biomass of trees by 7%. In this study, ambient $[O_3]$ influenced both above and below ground productivity similarly. Compared with CF controls elevated $[O_3]$ of 97 ppb decreased total biomass of trees by 17% whereas elevated $[O_3]$ of 64 ppb reduced total biomass by 11% than the trees grown at ambient $[O_3]$. Further, the higher sensitivity of root biomass to elevated $[O_3]$ was suggested from the reduced root-to-shoot ratio at elevated $[O_3]$. At elevated $[O_3]$, trees exhibited significant declines in LA, Rubisco content and chlorophyll content which may bring about significant reductions in photosynthesis. When grown at elevated $[O_3]$ trees were of shorter height and reduced diameter with lower transpiration rates. In addition to this gymnosperms, were significantly less sensitive to elevated $[O_3]$ than angiosperms. But, few observations of the interaction of $[O_3]$ with elevated $[CO_2]$ and drought to convincingly reveal how these climate change factors will modify tree responses to $[O_3]$. Taken together, these results reveal that the carbon-sink strength of northern hemisphere forests is likely to decline by current $[O_3]$ and will be further decreased in the future if $[O_3]$ rises. This proposes that an important portion of global fossil-fuel CO_2 emissions is currently counteracted by a key carbon sink and could be condensed or lost in the future.

23.11 MODELING OF TREES

23.11.1 TECHNIQUE USED FOR MAPPING URBAN TREE

Xiao et al., (2004) mapped urban forest tree species using AVIRIS data and multiple-masking techniques. They mention that tree type and species information are vital parameters for urban forest management, benefit-cost analysis, and urban planning. But, conventionally, these parameters have been found from limited field samples in urban forest management practice. For recognition and mapping of urban forest, trees high-resolution airborne visible infrared imaging spectrometer (AVIRIS) data and multiple-spectral masking techniques are employed in this study. Trees were identified on the basis of their spectral character difference in AVIRIS data. The confounding noise during spectral analysis can be decreased by directing the attention to the target land cover types only by using multiple-masking techniques. Ground reference data and comparison of tree information in an existing geographical information system (GIS) database were utilized to verify the results. The difference in both tree type and species was exhibited from their different mapping accuracy which was 94% at the tree type level and 70% at the tree species level. For the 12 deciduous tree species, the average accuracy obtained was 70% while for four evergreen tree species, the average accuracy was 69% and the comparatively low accuracy for several deciduous species was because of their small tree size and overlapping among tree crowns at the 3.5 m spatial resolution of AVIRIS data. Thus, this urban forest tree species mapping method reduces the costs of mapping compared to field sampling or other traditional methods and offers the possibility of increasing data update intervals and precision.

23.11.2   *MULTIPLE DECISION TREE MODELS*

Tong et al. (2003) combined the outcomes of multiple classification models to produce a single prediction for many years and developed decision tree models. In earlier applications, the multiple models to be combined were created by altering the set. The risk of decreasing predictivity of the individual models to be combined and/or over fitting the noise in the data may be caused by the usage of these so-called resampling techniques which might give poorer prediction of the composite model than the individual models. A new method called Decision Forest which combines multiple Decision Tree models is proposed by the authors in this study. Each Decision Tree model developed using a unique set of descriptors. The quality of individual models is constantly and significantly enhanced by combining the models of related predictive quality using Decision Forest method in both training and testing steps.

23.12 MISCELLANEOUS TECHNIQUES

23.12.1 CARBON SEQUESTRATION

Goulden et al. (1996) used long-term eddy covariance methods and a critical estimation of accuracy for measuring the carbon sequestration. Using the eddy covariance technique, the turbulent exchanges of CO_2 and water vapor between an aggrading deciduous forest in the north-eastern the United States (Harvard Forest) and the atmosphere from 1990 to 1994 was measured. The comprehensive explanation of the approaches used and a rough estimate of the precision and accuracy of these measurements were presented. Analysis of the surface energy budget reveals a uniform systematic error in the turbulent exchange measurements of -20 to 0%. A selective systematic underestimation during calm (friction velocity < 0.17 m s^{-1}) nocturnal periods was unveiled from a comparison of nocturnal eddy flux with chamber measurements and for correcting this error a method was described. As a result of sampling uncertainty of ± 0.3 t C ha^{-1} y^{-1} determined by Monte Carlo simulation in 1994, the integrated carbon sequestration was 2.1 t C ha^{-1} y^{-1} with a 90% confidence interval. When the direct observations are not available, sampling uncertainty may be reduced by evaluating the flux as a result of the physical environment by minimizing the length of intervals without flux data. These analyses lead us to place an overall uncertainty on the annual carbon sequestration in 1994 of -0.3 to $+ 0.8$ t C $ha^{-1}y^{-1}$.

Lal and Kimble (1997) examined conservation tillage for carbon sequestration compared with about 700 Pg in the atmosphere and 600 Pg in land biota world soils function as the largest terrestrial sink of organic carbon (C), about 1550 Pg. The atmospheric carbon pool increases significantly due to different agricultural activities such as, deforestation, burning, plowing and intensive grazing. Further agriculture expansion might have added considerably to the atmospheric carbon pool. But the precise magnitudes of carbon fluxes from soil to the atmosphere and from land biota to the soil are not known. An important goal of the sustainable management of soil resources is to augment soil organic carbon (SOC) pool by improving passive or non-labile fraction. Soil surface management, soil water conservation and management, and soil fertility regulation are all important aspects of carbon sequestration in soil. All tillage methods that decrease runoff and soil erosion in comparison with plow-based tillage known to improve SOC content of the surface layer is termed as conservation tillage. Increase in micro-aggregation and deep placement of SOC in the sub-soil horizons are the primary mechanisms of carbon sequestration with conservation tillage. Enhanced biomass production (e.g., soil fertility enhancement, improved crops and species, cover crops and fallowing, improved pastures and deep-rooted crops) are other beneficial agricultural practices related with conservation tillage. Furthermore, it is suitable to implement soil and crop management practices that put emphasis on humification and improve the passive fraction of SOC. Owing to the importance of C sequestration, soil quality should be estimated in terms of its SOC content.

Batjes and Sombroek (1997) reviewed potentials for carbon sequestration in tropical and subtropical soils. As soil, organic matter (OM) is a key constituent of all terrestrial ecosystems, any variation in its composition and abundance has significant influences on many of the activities that occur within the system. In the background of agricultural sustainability and global environmental change, the function of soil OM in soil nutrient cycling and soil gaseous emissions was discussed. With specific reference to the subtropical and tropical regions, data on organic carbon and nitrogen reserves in the soils of the world were presented and potentials for long-lasting, augmented sequestration of carbon in the soil through the management of the land and water resources are studied. Conclusively, to permit the assessment of variations in soil OM quantity and quality over time, as verified by changes in land-use and climate, the necessity is emphasized for an up to date database on soil resources and for a global monitoring system.

Oren et al. (2001) stated soil fertility boundaries to carbon sequestration by forest ecosystems in a CO_2-enriched atmosphere. They reveal that northern mid-latitude forests are a huge terrestrial carbon sinks. Regardless of nutrient restrictions, large enrichment in carbon sequestration from carbon dioxide (CO_2) fertilization is estimated in these forests. However, forests are usually considered as assets of adequate to poor fertility, where tree growth is frequently restricted by nutrient supply, in specific to nitrogen. The authors presented evidence that estimates of increases in carbon sequestration of forests, which is expected to partially compensate for increase CO_2 in the atmosphere, are unduly optimistic (Körner, 1995). The CO_2-induced biomass carbon addition devoid of added nutrients was not evident at a nutritionally poor site and the stimulation at a nutritionally moderate site was transitory, stabilizing at a marginal gain after three years were observed in two forest trials on maturing pines subjected to raise atmospheric CO_2. But an enormous synergistic advance from higher CO_2 and nutrients was identified with nutrients added. Compared with the moderate site (twofold higher), this advancement was larger at the poor site (threefold greater than the expected additive effect). Hence, the response of wood carbon sequestration to augmented atmospheric CO_2 can be controlled by fertility. Evaluation of future carbon sequestration has to study the restrictions enacted by soil fertility, accompanied by interactions with nitrogen deposition.

Lal (2005) reviewed forest soil carbon sequestration. High carbon (C) density is seen in soils having stability with a natural forest ecosystem. The soil C stock gets decrease with land use change, chiefly conversion to agricultural ecosystems and with altitude the ratio of soil:vegetation C density increases. Thus pre degraded agricultural soils have less SOC stock than their probable ability. So SOC stock through C sequestration can be improved by afforestation of agricultural soils and management of forest plantations. Between climates, soils, tree species and management and chemical composition of the litter, the quantity of SOC sequestration and the magnitude and quality of soil C stock gets affected and reliant on the complex interaction as revealed by the dominant tree species. Increasing the production of forest biomass per se may not essentially improve the SOC stocks. Soil C stock can get affected for a long period with fire natural or managed which is an important disturbance. By a careful site preparation, sufficient soil drainage, growing species with a high NPP, application of N and micronutrients (Fe) as fertilizers or biosolids and preserving soil and water resources can significantly enhance the soil C stock. Further, by the raising availability of mineral N and via CO_2, fertilization effect climate change could also improve forest growth which may partially counteract the release of soil C in response to warming. In the measurement of soil C stock and fluxes and scaling of C stock from pedon/plot scale to regional and national scales important developments have been achieved. To ameliorate variations in atmospheric chemistry soil C sequestration in boreal and temperate forests may be implemented as a significant approach.

Blain et al. (2007) investigated the natural iron fertilization effect on carbon sequestration in the Southern Ocean. They mention that the accessibility of iron restricts primary productivity and the related carbon uptake over large areas of the ocean. Any change in iron supply to the surface ocean may have a significant influence on atmospheric carbon dioxide concentrations over glacial-interglacial cycles as iron performs an essential function in the carbon cycle and currently, the function of iron in carbon cycling has mainly been verified using short-term iron-addition experiments. But it is hard to reliably determine the extent of carbon export to the ocean interior using such methods and extrapolation of the results to longer timescale may be involved in the short observational periods. To overcome some of the confines of short-term experiments, observations of a phytoplankton bloom produced by natural iron fertilization is a method that proposes the opportunity. They observed that a large phytoplankton bloom over the Kerguelen Plateau in the Southern Ocean was upheld by the supply of iron and major nutrients to surface waters from iron-rich deep water below. From short-term blooms brought by iron-addition experiments, the ratio of the carbon export to the amount of iron supplied, i.e., the efficiency of fertilization was observed to be at least ten times higher than previous assessments. This consequence gives new vision on the influence of long-term fertilization by iron and macronutrients on carbon sequestration, indicating that deviations in iron supply from below is raised in some palaeoclimatic and future climate change situations may have a further important effect on atmospheric carbon dioxide concentrations than previously thought.

Batjes and Sombroek (1997) reviewed the experimental evidence for long-term carbon (C) sequestration in soils as significance of specific forest management strategies. Through the incorporation of carbon dioxide (CO_2) in plant biomass the land-use changes for instance which result from afforestation and management of fast-growing tree species, have an instant influence on the regional rate of C sequestration and owing to the environmental and political constraints the possibilities of such practices is restricted in Europe. The management of current forests can also increase C sequestration, however, earlier the authors considered the consequences of harvesting, thinning, fertilization application, drainage, tree species selection, and control of natural disturbances on soil C dynamics. The aspects that affect the C input to the soil and the C release through decomposition of soil organic matter (SOM) was focused. Smaller amount information is accessible about quantifiable influences of management on stable C pools in the mineral soil but sufficient evidences are there regarding the influences of management on the extent of C in the organic layers of the forest floor. The distinction of SOM into labile and stable soil C segments is essential. The risk of accidental C losses can be prevented by reducing the disturbances in the stand structure and soil and the stable pool C storage capacity can be enlarged by improving the productivity of the forest and thus improving the C input to the soil. Further, the stability of the forest can be enhanced by establishing mixed species which can circumvent high rates of SOM decomposition and between tree species; the amount of C accumulation and its allocation within the soil profile differs. However, variances in the stability of SOM as a direct species effect have not yet been reported.

23.12.2   SAP FLOW OF WOOD

Burgess et al. (2001) explained an improved heat pulse method to measure low and reverse rates of sap flow in woody plants. They state that because of the limitations of the compensation heat pulse method (CHPM) it is of low value for measuring low rates of sap flow in woody plants. So in this study, an advanced heat pulse method termed the heat ratio method (HRM) is presented for determining low and reverse rates of sap flow in woody plants. Improved measurement range and resolution, protocols to correct for physical and thermal errors in sensor deployment and a simple linear function to define wound effects are few significant benefits of HRM over the CHPM. The theory and methodological protocols of the HRM against gravimetric measurements of transpiration that offer wound correction coefficients and validate the consistency and precision of the technique are described.

Čermák et al. (2004) conducted a study on sap flow measurement techniques using some thermodynamic methods flow integration within trees and scaling up from sample trees to entire forest stands. Based on 30 years' experience the trunk segment heat balance (THB) and heat field deformation (HFD) methods were given a specific consideration. Considering the difference of radial patterns in sapwood and difference around stems expansion of sap flow data in terms of integrating flow for whole stems from individual determining points were discussed. Further, with a discussion of the magnitude of errors scaling up of data from sets of example trees to the whole forest stands based on broadly accessible biometric data (partially on remote sensing images) is explained and measured, the routine method applicable in any forest stand and applied examples.

23.13   CHEMICAL COMPOSITION

23.13.1   FODDER

Larbi et al. (1998) examined some multipurpose fodder trees and shrubs for their chemical composition, rumen degradation, and gas production characteristics during wet and dry seasons in the humid tropics. For 18 multipurpose trees and shrubs (MPTs) from the humid lowlands of West Africa, seasonal variations in chemical composition, dry matter (DM) and nitrogen (N) degradation and gas production features were analyzed. The MPTs add for the improvement of integrated crop and livestock agroforestry technologies in the region. During the main-wet (April-August) and dry (December-March) seasons, the research was initiated in Ibadan, southwestern Nigeria. The degradation and gas production characteristics were found to be connected to chemical composition based on which the MPTs are ordered. Extensive variances were not noticed among MPTs in crude protein (CP), neutral detergent fiber (NDF), acid detergent fiber (ADF) and proanthocyanidin contents, DM, and N degradation and gas production characteristics. During the dry season, DM degradation varied from 416 to 868 g kg^{-1} and for N 508 to 950 g kg^{-1}. The rates and extents of DM and N degradation were significantly associated with NDF and ADF during the wet season ($r = -0.47$ to -0.63). The volume of gas produced ($r = -0.48$ to -0.67) and initial gas production ($r = -0.64$ to -0.73) were highly correlated with the NDF and ADF in both seasons. In the minor-wet, season the proportion of DM degradation exhibited significant association with gas production variables. Volume of gas produced for the main-wet and dry seasons and classification of the MPTs on the basis of the magnitude of DM and N degradation exhibited significant associations. In the main-wet and the dry seasons *F. exasperata, S. nodosa, S. siamea, S. spectabilis, G. sepium, L. leucocephala,* and *L. diversifolia* were found to be superior in quality than *M. thonningii, A. angustissima,* and *P. pterocarpum* based on their degradation and gas production characteristics.

El Hassan et al. (2000) compared the alfalfa (*Medicago sativa*) hay and teff (*Eragrostis abyssinica*) straw with samples of foliage from multipurpose leguminous trees (MPT), chosen as potential feed supplements for ruminants for their chemical composition and *in situ* degradation traits. They verified OM, ADF, NDF, nitrogen, neutral detergent nitrogen, acid detergent lignin (ADL), soluble phenolics, NDF-bound proanthocyanidins and in addition to this *in vitro* digestibility in *Acacia angustissima, Chamaecytisus palmensis* (Tagasaste), *Leucaena leucocephala,* two cultivars of *Sesbania sesban* and *Vernonia amygdalina* (bitter leaf) were analyzed. The teff straw contains a high fiber, but low N (4.0 g available N per kg DM) which can be added from MPT containing higher quantities of nutrient mainly N (up to 39.5 g available N per kg DM) comparable to alfalfa hay. Besides, they carried out *in situ* nylon bag digestion and *in vitro* gas production analyses to determine microbial degradation characteristics. As the MPT:teff straw ratio increases in *A. angustissima, in vitro* gas production decreases though the MPT were highly degradable *in situ*. This unveils that antimicrobial components were found in this species. None of the chemical estimations were correlated with antimicrobial properties. Ultimately it is determined that for ruminants receiving poor quality forages like teff straw some of the MPT verified may be suitable dietary supplements. But for new MPT containing antimicrobial components or material toxic to the animal itself only chemical analysis will be of less value in predicting the nutritive value.

23.13.2 CHEMICAL COMPOSITION OF WOOD

Pettersen (1984) considered the chemical composition of wood including overall chemical composition of wood, methods of analysis, structure of hemicellulose components and the degree of polymerization of carbohydrates. From several countries, wood data are collected in tables. From 1927–68 data were collected at Forest Products Laboratory (Madison, Wisconsin), which acquired from both United States and foreign woods. Data pertaining to compositions of woods from Borneo, Brazil, Cambodia, Chile, Colombia, Costa Rica, Ghana, Japan, Mexico, Mozambique, Papua New Guinea, the Philippines, Puerto Rico, Taiwan, and the USSR was included in previously published data. Different wood components such as cellulose (Cross and Bevan, holo-, and alpha-), lignin, pentosans, and ash were analyzed and solubilities in 1% sodium hydroxide, hot water, ethanol/benzene, and ether are documented. For common temperate-zone, woods include the individual sugar composition (as glucan, xylan, galactan, arabinan, and mannan), uronic anhydride; acetyl, lignin, and ash. More detailed analyses data were presented.

Curling et al. (2002) examined associations between mechanical properties, weight loss, and chemical composition of wood during incipient brown-rot decay. Before measurable loss occurs in wood, incipient decay of wood by brown-rot fungi decreases strength losses. From the previous report, it is found that loss in hemicellulose may be the cause for high levels of strength loss during incipient brown-rot decay. Therefore, the association of decay with the mechanical properties of the wood and its influence on hemicellulose composition was analyzed with an *in vitro* test method which allowed progressive sampling of southern pine subjected to monocultures of brown-rot fungi. Using mechanical testing and chemical analysis, the wood was evaluated consequently. About 4:1 ratio of strength to weight loss was unveiled from the results. Additionally, it is found that the early strength loss (up to 40%) associated with loss of arabinan and galactan components and following strength loss (greater than 40%) was related with the loss of the mannan and xylan components. Significant loss of glucan (representing cellulose) was only noticed at greater than 75% modulus of rupture.

Baldock and Smernik (2002) analyzed the thermally altered *Pinus resinosa* (Red pine) wood for chemical composition and bioavailability. They mention that the residues left after incomplete combustion of vegetation (char) can significantly increase the soils carbon content. The quantity of biological inertness probably rests on the extent that the combustion residues were transformed during thermal treatment and compared to other forms of organic C in soils Char C is considered biologically inert. By heating *Pinus resinosa* sapwood to temperatures between 70 and 350°C in the laboratory, char C was produced and the relationship between deviations in chemical composition and biological inertness was measured. Till the mass of the residual char material is stabilized, (± 2%) heating at each temperature was continued. Using elemental analysis, solid-state ^{13}C nuclear magnetic resonance (NMR) spectroscopy and diffuse reflectance infrared Fourier transform spectroscopy (DRIFT), the chemical composition of char was verified. At temperatures ≥200°C thermal treatment produced significant variations in chemical composition and in a 120-day laboratory, incubation liability of the heated sapwood to biological oxidation was enumerated. With a preliminary dehydration and the formation of unsaturated structures variations in elemental substances and molar elemental ratios were found to be constant. A conversion of O-alkyl C to aryl and O-aryl furan-like structures occurs, owing to the changes in the chemical composition with increasing heating temperature which was unveiled from the NMR and DRIFT data, and found to be constant with results from work observing the chemical changes related with thermal treatment of the main constituent of wood cellulose. The ability of microbial inoculums derived from decomposing *Pinus resinosa* wood to mineralize carbon in the charred samples, was declined significantly with these chemical changes. Further, by an order of magnitude for wood heated to $\geq 200°C$, the C mineralization rate constants were decreased.

Ferraz et al. (2000) assessed the chemical composition of biodegraded pine and eucalyptus wood by DRIFT spectroscopy and multivariate analysis. For monitoring wood biodegradation Fourier transformed infra-red (FTIR) was applied as an analytical tool. Six white- and two brown-rot fungi was used for decaying a sample set with typical soft (*Pinus radiata*) and hardwood (*Eucalyptus globulus*). Provided samples underwent weight-loss differing from 0.4% to 36% for pinewood and 1.7% to 42% for eucalyptus wood within 30 days to 1 year biodegradation times. Decayed samples were characterized by conventional wet chemical analysis and by diffuse reflectance FTIR (DRIFT) spectroscopy. Further, to relate chemical composition in wood samples with the FTIR spectral data, multivariate analysis was used. In most cases, the major wood components, concentrations at the 99% confidence level presenting r^2 values higher than 0.86 were predicted from partial least squares (PLS) models. Models for *P. radiata* were more accurate than for *E. globulus*.

23.14 CONCLUSIONS

Concerted research inputs have been directed on various aspects of technology development of ecophysiological traits and biochemistry of plants and trees. This involves morphology, anatomy, general ecophysiology, specific techniques, and formation of rings, growth, and assessment of productivity, miscellaneous techniques, and chemical composition of woods. This information will serve as a guide to students and researchers to work on trees.

KEYWORDS

- biochemistry
- ecophysiology
- growth
- photosynthesis
- productivity
- respiration
- technology development
- trees

REFERENCES

Abrams, M. D., (1990). Adaptations and responses to drought in *Quercus* species of North America. *Tree Physiol., 7*, 227–238.

Abrams, M. D., (1994). Genotypic and phenotypic variation as stress adaptations in temperate tree species: A review of several case studies. *Tree Physiol., 14*, 833–842.

Abrams, M. D., & Mostoller, S. A., (1995). Gas exchange, leaf structure, and nitrogen in contrasting successional tree species growing in open and understory sites during a drought. *Tree Physiol., 15*, 361–370.

Abrams, M. D., Ruffner, C. M., & Morgan, T. A., (1998). Tree-ring responses to drought across species and contrasting sites in the ridge and valley of central Pennsylvania. *For. Sci., 44*, 550–558.

Ackerly, D., Knight, C., Weiss, S., Barton, K., & Stamer, K., (2002). Leaf size, specific leaf area, and microhabitat distribution of chaparral woody plants: Contrasting patterns in species level and community level analysis. *Oecologia, 130*, 449–457.

Baldocchi, D., (1997). Measuring and modeling carbon dioxide and water vapor exchange over a temperate broad-leaved forest during the 1995 summer drought. *Plant Cell Env., 20*, 1108–1122.

Baldock, J. A., & Smernik, R. J., (2002). Chemical composition and bioavailability of thermally altered *Pinus resinosa* (Red pine) wood. *Org. Geochem., 33*, 1093–1109.

Batjes, N. H., & Sombroek, W. G., (1997). Possibilities for carbon sequestration in tropical and subtropical soils. *Global Change Biol., 3*, 161–173.

Blain, S., Quéguiner, B., & Armand, L., et al., (2007). Effect of natural iron fertilization on carbon sequestration in the Southern Ocean. *Nature, 446*, 1070–1074.

Brooks, J. R., Flanagan, L. B., & Ehleringer, J. R., (1998). Responses of boreal conifers to climate fluctuations: Indications from tree-ring widths and carbon isotope analyses. *Can., J. For. Res., 28*, 524–533.

Burgess, S. S. O., Adams, M. A., & Turner, N. C., et al., (2001). An improved heat pulse method to measure low and reverse rates of sap flow in woody plants. *Tree Physiol., 21*, 589–598.

Büntgen, E., Frank, D. C., Kaczka, R. J., Verstege, A., Zwijacz-Kozica, T., & Esper, J., (2007). Growth responses to climate in a multi-species tree-ring network in the Western Carpathian Tatra Mountains, Poland, and Slovakia. *Tree Physiol., 27,* 689–702.

Canham, C., (1988). Growth and canopy architecture of shade-tolerant trees: Response to canopy gaps. *Ecology, 69,* 786–795.

Carrer, M., & Urbinati, C., (2004). Age-dependent tree ring growth responses to climate in *Larix decidua* and *Pinus cembra. Ecology, 85,* 730–740.

Čermák, J., Kučera, J., & Nadezhdina, N., (2004). Sap flow measurements with some thermodynamic methods, flow integration within trees and scaling up from sample trees to entire forest stands. *Trees, 18,* 529–546.

Ceulemans R., & Mousseau, M., (1994). Effects of elevated atmospheric CO_2 on woody plants. *New Phytol., 127,* 425–446.

Cimato, A., Castelli, S., Tattini, M., & Traversi, M. L., (2010). An ecophysiological analysis of salinity tolerance in olive. *Env. Exp. Bot., 68,* 214–221.

Close, D. C., Ruthrof, K. X., Turner, S., Rokich, D. P., & Dixon, K. W., (2009). Ecophysiology of species with distinct leaf morphologies: Effects of plastic and shade-cloth tree guards. *Rest. Ecol., 17,* 33–41.

Corcuera, L., Camarero, J. J., & Gil-Pelegrín, E., (2002). Functional groups in *Quercus* species derived from the analysis of pressure-volume curves. *Trees, 16,* 465–472.

Curling, S. F., Clausen, C. A., & Winandy, J. E., (2002). Relationships between mechanical properties, weight loss, and chemical composition of wood during incipient brown-rot decay. *For. Prod. J., 52,* 34–39.

Dawson, T. E., (1996). Determining water use by trees and forests from isotopic, energy balance and transpiration analyses: The roles of tree size and hydraulic lift. *Tree Physiol., 16,* 263–272.

DeJong, T. M., Da Silva, D., Vos, J., & Escobar-Gutiérrez, A. J., (2011). Using functional–structural plant models to study understand and integrate plant development and ecophysiology. *Ann. Bot., 108,* 987–989.

de Reffye, Ph., Blaise, F., & Houllier, F., (1998). Modeling plant growth and architecture: Some recent advances and applications to agronomy and forestry. *Acta Hortic., 456,* 105–116

Donovan, L. A., & Ehleringer, J. R., (1991). Ecophysiological differences among juvenile and reproductive plants of several woody species. *Oecologia, 86,* 594–597.

Eamus, D., & Prior, L., (2001). Ecophysiology of trees of seasonally dry tropics: Comparisons among phenologies. *Adv. Ecological Res., 32,* 113–197.

El Hassan, S. M., Lahlou-Kassi, A., Newbold, C. J., & Wallace, R. J., (2000). Chemical composition and degradation characteristics of foliage of some African multipurpose trees. *Anim. Feed Sci. Technol., 86,* 27–37.

Ferraz, A., Baeza, J., Rodríguez, J., & Freer, J., (2000). Estimating the chemical composition of biodegraded pine and eucalyptus wood by DRIFT spectroscopy and multivariate analysis. *Biores. Technol., 74*(3), 201–212.

Gale, M. R., & Grigal, D. F., (1987). Vertical root distributions of northern tree species in relation to successional status. *Can. J. Forest Res., 17,* 829–834.

Gebre, G. M., Tschaplinski, T. J., & Shirshac, T. L., (1998). Water relations of several hardwood species in response to through fall manipulation in an upland oak forest during a wet year. *Tree Physiol., 18,* 299–305.

Goulden, M. L., Munger, J. W., Fan, S. M., Daube, B. C., & Wofsy, S. C., (1996). Measurements of carbon sequestration by long-term eddy covariance: Methods and a critical evaluation of accuracy. *Global Change Biol.*, *2*, 169–182.

Gunderson, C. A., & Wullschleger, S. D., (1994). Photosynthetic acclimation in trees to rising atmospheric CO_2: A broader perspective. *Photosynth. Res.*, *39*(3), 369–388.

Hättenschwiler, S., Schweingruber, F. H., & Körner, Ch., (1996). Tree ring responses to elevated. CO_2 and increased N deposition in *Picea abies*. *Plant Cell Env.*, *19*, 1369–1378.

Kolb, T. E., & Stone, J. E., (2000). Differences in leaf gas exchange and water relations among species and tree sizes in an Arizona pine-oak forest. *Tree Physiol.*, *20*, 1–12.

Körner, C., (1995). Towards a better experimental basis for up scaling plant responses to elevated CO_2 and climate warming. *Plant Cell Env.*, *19*, 1101–1110.

Küppers, M., (1989). Ecological significance of aboveground architectural patterns in woody plants: A question of cost-benefit relationships. *Trends Ecol. Evol.*, *4*, 375–379.

Kubiske, M. E., & Abrams, M. D., (1994). Ecophysiological analysis of woody species in contrasting temperate communities during wet and dry years. *Oecologia*, *98*, 303–312.

Lal, R., & Kimble, J. M., (1997). Conservation tillage for carbon sequestration. *Nutrient Cycling in Agroecosystems*, *49*, 243–253.

Lal, R., (2005). Forest soils and carbon sequestration. *For. Ecol. Manage.*, *220*, 242–258.

Larbi, A., Smith, J. W., Kurdi, I. O., Raji, A. M., & Ladipo, D. O., (1998). Chemical composition, rumen degradation, and gas production characteristics of some multipurpose fodder trees and shrubs during wet and dry seasons in the humid tropics. *Anim. Feed Sci. Technol.*, *72*(1/2), 81–96.

Maiti, R. K., Gonzalez Rodriguez, H., Sarkar, N. C., & Thakur, A.K., (2015). Branching pattern and leaf crown architecture of some tree and shrubs in northeastern Mexico. *Intern. J. Biores. Stress Manage.*, *6*, 41–50.

Maiti, R., González Rodríguez, H., & Kumari, Ch. A., (2016). Applied biology of woody plants. American Academic Press. Salt Lake City, Utah, USA. 367 pp.

Newman, G. S., Arthur, M. A. & Muller, R. N., (2006). Above- and belowground net primary production in a temperate mixed deciduous forest. *Ecosystems*, *9*, 317–329.

Oren, R., Ellsworth, D., & Johnsen, K. et al., (2001). Soil fertility limits carbon sequestration by forest ecosystems in a CO_2-enriched atmosphere. *Nature*, *411*, 469–472.

Pepin, S., Plamondon, A. P., & Britel, A., (2002). Water relations of black spruce trees on a peatland during wet and dry years. *Wetlands*, *22*, 225.

Pettersen, R. C., (1984). The Chemical Composition of Wood. U.S. *The Chemistry of Solid Wood* (Chapter 2, pp. 57–126), Department of Agriculture, Forest Service, Forest Products Laboratory, Madison, WI 53705.

Potvin, C., Lechowicz, M. J., & Tardif, S., (1990). The statistical analysis of ecophysiological response curves obtained from experiments involving repeated measures. *Ecology*, *71*, 1389–1440.

Rascher, U., Liebig, M., & Luttge, U., (2000). Evaluation of instant light-response curves of chlorophyll fluorescence parameters obtained with a portable chlorophyll fluorometer on site in the field. *Plant Cell Env.*, *23*, 1397–1405.

Ritchie, G. A. (1984). Chapter assessing seedling quality. *Forestry Nursery Manual: Production of Bare Root Seedlings,* pp. 243–259, Part of the Forestry Sciences book series (FOSC, vol. 11).

Rodriguez, H. G., Maiti, R., & Kumari, C. A., (2016). Biodiversity of leaf traits in woody plant species in northeastern Mexico: A synthesis. *Forest Res. Open Access,* 5, 2.

Rozas, V., (2005). Dendrochronology of pedunculate oak (*Quercus robur* L.) in an old-growth pollarded woodland in northern Spain: Tree-ring growth responses to climate. *Ann. For. Sci.,* 62, 209–218.

Sack, L., Cowan, P. D., Jaikumar, N., & Holbrook, N. M., (2003). The hydrology of leaves: Co-ordination of structure and function in temperate woody species. *Plant Cell Env.,* 26, 1343–1356.

Siefert, A., Violle, C., Chalmandrier, L., & Albert, C. H., et al., (2015). A global meta-analysis of the relative extent of intraspecific trait variation in plant communities. *Ecol. Lett.,* 18, 1406–1419.

Sprugel, D. C., (1990). Components of woody-tissue respiration in young *Abies amabilis* (Dougl.) Forbes trees. *Trees,* 4, 88–98.

Tong, W., Hong, H., Fang, H., Xie, Q., & Perkins, R., (2003). Decision forest: Combining the predictions of multiple independent decision tree models. *J. Chem. Inf. Comput. Sci.,* 43, 525–531.

Vieira, J., Campelo, F., & Nabais, C., (2009). Age-dependent responses of tree-ring growth and intra-annual density fluctuations of *Pinus pinaster* to Mediterranean climate. *Trees,* 23(2), 257–265.

Wen, D., Kuang, Y., & Zhou, G., (2004). Sensitivity analyses of woody species exposed to air pollution based on ecophysiological measurements. *Environ. Sci. Poll. Res.,* 11, 165–170.

Wittig, V. E., Ainsworth, E. A., Naidu, S. L., Karnosky, D. F., & Long, S. P., (2009). Quantifying the impact of current and future tropospheric ozone on tree biomass, growth, physiology, and biochemistry: A quantitative meta-analysis. *Global Change Biol.,* 15, 396–424.

Wright, I. J., Reich, P. B., Cornelissen, J. H. C., et al. (2005). Assessing the generality of global leaf trait relationships. *New Phytol.,* 166, 485–496.

Xiao, Q., Ustin, S. L., & McPherson, E. G., (2004). Using AVIRIS data and multiple-masking techniques to map urban forest tree species. *Intern. J. Remote Sens.,* 25, 5637–5654.

Yan, G., & Baoguo, L., (2001). New advances in virtual plant research. *Chinese Science Bulletin.* 46, 888; doi: 10.1007/BF02900459.

Zeppel, M. J. B., Macinnis-Ng, C. M. O., Yunusa, I. A. M., Whitley, R. J., & Eamus, D., (2008). Long term trends of stand transpiration in a remnant forest during wet and dry years. *J. Hydrol.,* 349, 200–213.

PART IV
RESEARCH ADVANCES

CHAPTER 24

Modern Methods and Research Directions of Biochemistry of Boreal Woody Plants

EKATERINA SERGEEVNA ZOLOTOVA[1] and
NATALYA SERGEEVNA IVANOVA[2,3]

[1]*Zavaritsky Institute of Geology and Geochemistry, Ural Branch,
Russian Academy of Science, 15 Akademika Vonsovskogo Street,
Yekaterinburg – 620016, Russia*

[2]*Ural State Forest Engineering University, Yekaterinburg,
37, Sibirskiy trakt, Yekaterinburg – 620100, Russia*

[3]*Botanical Garden of the Ural Branch of the
Russian Academy of Sciences, Yekaterinburg, 202a,
8th March Street Yekaterinburg – 620144, Russia*

24.1 ACCUMULATION AND DISTRIBUTION OF MICRO- AND MACRO ELEMENTS IN WOODY PLANTS

The exchange of chemical substances in plant communities has been the subject of specialized studies, which are aimed at elucidating the characteristics of mineral nutrition of plants. The absorption and transformation of mineral substances is the basis for regulating the processes of synthesis of organic matter, and plants' growth. Plants are characterized by their selective ability to absorb certain elements from the soil, which causes the accumulation of such important elements of fertility as N, P, K, etc. (Vinokurova and Lobanova, 2011).

Mineral substances accumulated in plants are divided into macro-elements (their share of the dry matter of plants is higher) and micro-elements (less than 0.001%) (Lugansky et al., 2010).

The specificity of accumulation and distribution of macro-elements in various parts of the basic woody species of spruce and fir plantations were observed in the Mari El Republic. Middle-aged model trees of *Picea abies, Abies sibirica, Tilia cordata,* and *Betula pendula* were selected as research objects. The list of the minimum values of the ash elements content in the needles of the current year of *Picea abies* and *Abies sibirica* trees is noted. These values are sufficient for optimum nutrition in the middle-aged spruce and fir plantations. It is observed that hardwoods such as *Tilia cordata* and *Betula pendula* are good companions of the conifers. In this respect, the leaf fall of the trees which contain a significant amount of nitrogen, potassium and phosphorus improves soil fertility of the spruce and fir forests (Vinokurova and Lobanova, 2011).

The individual organs of each tree species was analyzed for the contents of macro elements in the following manner (Vinokurova and Lobanova, 2011):

Picea abies and *Abies sibirica*: needles > branches > bark > roots > wood;
Betula pendula: leaves > roots > bark > branches > wood;
Tilia cordata: leaves > branches > roots > bark > wood.

The mineral elements content in xylem sap of *Pinus sylvestris* L. (Sazonova et al., 2009), wood (Demakov et al., 2013) and of the dry needles (Khabarova et al., 2015) were analyzed. Also we found the development of integrated indicators of functional diagnostics in the supply of pine seedlings with individual elements. For example, the defining optimal boron content in a plant according to the tolerance of the seedling to the snow blight (caused by *Phacidium infeslans* Karst.) (Yalynskaya and Chernobrovkina, 2008).

The elemental composition of xylem sap of *Pinus sylvestris* L. and its seasonal dynamic were studied in the Karelia Republic, depending on the vital status of the tree (dominant and suppressed in the canopy) and also from the forest type. The N, P, K contents and N:P:K ratio in xylem sap change regularly in the vegetation and are correlated with the intensity of growth. However, the mineral nutrient contents vary during vegetation more than N:P:K ratio. The mean values of the contents and the ratio of macro elements in the vegetation do not differ during the period of research. The independence was shown in the case of the N, P, K contents and N:P:K ratio in xylem sap from the metabolic status of the trees growing in the same stand. The dependence was found of the xylem sap composition (N, P, K, Ca, Mg, Mn, Zn, Cu, Al, Fe) on soil conditions in different forest types (Sazonova et al., 2009).

Demakov and his colleagues (2013) studied the mineral nutrition (according to the wood ash composition) of 13 tree species in a short-term flooding biotope. The gross annual consumption by the stands of the basic minerals and its total accumulation in the stem wood were calculated. The tree species that grow within a single floodplain biotope differ significantly in the effectiveness of their use of nutrients. Balsam poplar (*Populus balsamifera*) utilize more mineral substances in the production of 1 kg of timber. European white elm (*Ulmus laevis*) shows a little lower value. These two tree species most of all consume calcium and potassium. Siberian larch (*Larix sibirica*) is the one that uses soil potential most effectively (Demakov et al., 2013).

Pine needles contain the largest amount of ash elements and nitrogen compared to other parts of the tree. The ash content of the needles varies with age. The ash composition of the needles, which completes the life cycle is the most important source of replenishment of nutrients in the soils of forests. The ash content of the old needles differs significantly from the composition of the annual needles (Kazimirov and Morozova, 1973). The low content of any element in the leaves (needles) indicates an unsatisfactory supply of the trees with this element and is accompanied by inhibition of growth processes and as a result, by the decrease in forest productivity (Bobkova, 1987).

Habarova and her colleagues (Habarova et al., 2015) presented the results of the research they conducted on the influence of draining on the mineral element content of needles of Scotch pine. Growing and dead needles were inspected by wave-dispersive spectrometry for the content of the following elements: nitrogen, potassium, phosphorus, calcium, magnesium, sulfur, manganese, silicon, iron, aluminum, and sodium. The correlation ratio of element content and the distance from the dehydrator shows strong and significant correlation.

The research conducted by Tyukavina and Kunnikov (2015) relied on the X-ray fluorescence analysis of mineralogical composition of needles, branches, bark of Scotch pine and wood of Balsam poplar (*Populus balsamifera*) as possible components of the fertilizer composition. The percentage content of inorganic elements in pine biomass decreases in the following order: pine needles, branches of the 1st and the 2nd year, bark. The percentage levels of the mineral elements in the rotten wood of poplar and pine needles are the same. The content of ash constituent in poplar rotten wood is 2.1 times higher than in pine needles. The highest content of iron, calcium, potassium, silicon, magnesium and sodium is noted in the poplar rotten wood. The highest content of nitrogen, phosphorus, sulfur, manganese is observed in pine needles. Tree bark contains more calcium and aluminum than branches. Therefore, the combined use of the biomaterials will provide a more complete fertilizing composition, but it should be complemented by a source of phosphorus.

Annual tree rings are one of the most important carriers of information about the state of forest ecosystems (Shiyatov, 1986). Current techniques and methods only allow you to decipher a part of the information encoded in the annual rings of trees. The evaluation of the ash elements composition in annual rings, the content of which depends on the environmental conditions and the state of the trees is very promising (Adamenko et al., 1982). The study of the ash elements content is necessary both for understanding the process of their consumption by plants and for determining the regularities of the flow of the biological cycle in the ecosystem and for assessing the degree of environmental contamination when conducting ecological monitoring (Demakov and Shvetsov, 2013). Data describing the ash composition of annual layers of Scots pine (Demakov and Shvetsov, 2013) and Gmelin larch (Grachev et al., 2013, Vaganov et al., 2013) are given.

In periods of 20 years, the ash composition of annual layers of Scots pine in lakeside sites was studied. The content of Ca, Fe, Cu, Pb, Co and Cd in the wood of different annual layers varies synchronously in time: the concentration of the first three is steadily increasing, and the last three are fluctuating the minimum being the period from 1880 to 1940. The change in the content of other elements in the wood occurs in each isotope is extremely specific, but in general, it is also periodic. The dynamics of the content of the Zn, Sr, Cr and Ni in the annual layers of pines differ especially sharply (Demakov and Shvetsov, 2013).

In the case of the *Larix gmelinii* Rupr., changes in the content of the most important elements (P, K, Ca, Mg, Sr, Ba, Cl, Si) was studied from 1300 to 2000 ad period in the annual rings. The 700-year interval was provides samples of both living and very well-preserved dead trees of the Taimyr Peninsula. The chemical analysis was carried out by inductively coupled plasma mass spectrometry (ICP-MS) on quadrupole spectrometer "Agilent 7500ce." The measurements were made in media prepared by dissolving wood samples in concentrated nitric acid and then diluting them with water (Grachev et al., 2013). The analysis shows that different chemical elements (even if their functions in the cells life are similar) have different time variability. It means that they potentially contain valuable information about processes that are determined their inclusion in the xylem cell walls (Vaganov et al., 2013).

The influence of the urban environment on the changes of mineral exchange in woody plants was studied in Izhevsk (Buharina et al., 2007) and Ufa (Buharina and Dvoeglazova, 2010). The following methods were used in the studies: photocolorimetric method using the Nessler reagent (nitrogen content); Truogu-Meyeru method (phosphorus); flame photometry (potassium). The features of the distribution of basic elements of mineral nutrition in shoots of various tree species were being investigated. *Betula pendula, Acer negundo, Tilia cordata, Populus balsamifera, Sorbus aucuparia,* and *Caragana arborescens* investigated in Ufa. Dispersion analysis showed that the content of nitrogen, phosphorus and potassium in the plants depends on the species specific features, the conditions of the place of growth, the terms of vegetation of plants, the structural part of the plant and the interaction of these factors (P $<$ 10-18 $-$ 10-2). The studied species of woody plants differ substantially in the nitrogen content of the shoots. The content of the elements was calculated in percentage of absolutely dry mass (Buharina and Dvoeglazova, 2010).

The study of the features of mineral nutrition and the accumulation of elements in various organs of woody plants is necessary for understanding their physiology, assessing the growth and development of stands, studying their adaptation to natural and anthropogenic factors. Scientists used both time-proven methods and new ones, for example, inductively coupled plasma mass spectrometry (ICP-MS), X-ray fluorescence analysis. The work to study the characteristics of mineral nutrition is extremely impor-tant for solving the problem of increasing the productivity of forests.

24.2   SIGNIFICANT PHYSIOLOGICAL PROCESSES

24.2.1   PHOTOSYNTHESIS

Photosynthesis is a key link in the complex system of metabolism, ensuring the growth and development of plants according to the genetic program (Rubin, 2005).

Photosynthesis is one of the most sensitive processes to external conditions (Ilkun, 1978). The important factors are the intensity and spectral composition of the light flux, CO_2 content in the ambient air, temperature, humidity, soil moisture and its physico-chemical properties, wind, the intensity and direction of the atmospheric electric field, etc. (Karasev, 2001).

High-intensity light inhibits photosynthesis and causes destruction of pigment complexes (Ilkun, 1978; Karasev, 2001). Due to the mutual shading of the leaves, the photosynthetic apparatus of woody plants is much more complexly structured than that of herbs (Veretennikov, 1987; Buharina et al., 2007).

The parameters of the light curves change depending on the growing conditions. The light curve and data on the arrival of solar radiation on a cloudless day at a planned geographic location on a certain date of the vegetation period serve as the basis for constructing the «potential» curve (i.e. depending only on the light intensity) of photosynthesis. The light curves and data on the arrival of solar radiation are also the basis for calculating the maximum possible total absorption of CO_2 per 1 m² leaf surface per day. But in natural conditions, other external factors can be far from optimal so the real daytime dynamics of photosynthesis differ to a greater or lesser extent from that calculated from the light curve. For this reason, the dependence of photosynthesis should be obtained from all environmental factors (Molchanov, 2015).

A.G. Molchanov established that the rate of photosynthesis is dependent on the photosynthetic-active radiation reaching the leaf over the course of a day in different canopy layers of *Betula pendula* (Roth.) (Molchanov, 2015) and from the lack of moistening for *Quercus robur* (Molchanov, 2012).

For individual tree species, scientists have proved that the light curves of photosynthesis at different temperatures differ in the amount of light saturation, and the angle of tilt (Katrushenko, 1983). Zavyalova (1976) established the influence of the habitat on the light curves of photosynthesis: in the worst conditions, the curves acquire features characteristic of shady plants. At present, the fact of differences in assimilation parameters for light and shadow foliage (needles) is universally recognized (Molchanov, 2015).

Kholoptseva and her colleagues (2012) studied the effect of soil temperature, air temperature and light on the photosynthesis of silver birch (*Betula pendula*) seedlings. They conducted a multifactor experiment for two-year intact seedlings in a controlled environment. The greatest potential maximum of net-photosynthesis in silver birch seedlings was obtained at a soil temperature of 15.0°C, air of 20.6°C and an illumination of 34.6 lx. At the other two soil temperatures, the potential maximum of net-photosynthesis was lower, on average by 10 %, and was achieved at higher light and lower air temperature (Kholoptseva et al., 2012).

Bolondinskiy (2010) chose karelian birch and silver birch as the objects of research. The influence of light exposure, temperature and humidity on photosynthesis is investigated. The author used methods of mathematical modeling. The light and temperature optimum of photosynthesis, the combination of factors for maximal rates of photosynthesis were determined. It was shown that the absorption of CO_2 was higher by silver birch under atmospheric drought conditions.

Daylight photosynthesis could be used in photosynthetic productivity modeling taking into account the environmental factors of irradiance, water availability, and air temperature (Molchanov, 2015).

Many studies are devoted to the research of the dependence of photosynthesis and photosynthetic production of coniferous species from the influence of environmental factors. The following species were explored: *Pinus sylvestris* L. (Bolondinskiy and Vilikainen, 1987; Suvorova, 2009; Suvorova et al., 2012), *Picea obovata* Ledeb. (Suvorova, 2009; Suvorova et al., 2012), *Larix sibirica* Ledeb. (Suvorova, 2009; Suvorova et al., 2012) and *Picea exelsa Link* (Kurets and Drozdov, 2008).

Photosynthesis and mineral nutrition are closely related and interdependent processes. If mineral nutrition stimulates the formation of the photosynthetic apparatus and the intensity of its work, then intensive photosynthesis, in turn, is a condition conducive to the effective use of elements of mineral nutrition. Kramer and Kozlovsky (1983), Kholoptseva and Chernobrovkina (2009), Lebedev (2015), and Zarubina (2016) studied the influence of mineral nutrition on photosynthesis. The intensity of photosynthesis is reduced approximately 2-fold due to a lack of nitrogen in the soil. However, in the technogenic environment, there is often a reverse trend. An increase in the nitrogen content in leaves inhibits the synthesis of chlorophyll and adversely affects the photosynthetic activity of plants.

Zarubina (2016) researched the influence of nitrogen on the physiological and growth activity of young *Picea abies* L. in the bilberry birch forests (53- and 59-year-old) of the northern taiga. The potential (CO_2-saturated) photosynthesis and the outflow of ^{14}C-assimilates from spruce undergrowth were studied by the radiometric method with the specific radioactivity of the gas mixture ($^{14}CO_2$ + CO_2) in the chamber of 0.2 and 8 MBq/l and the duration of the experiment 10 and 30 min, respectively. Different doses of nitrogen have an ambiguous influence on young spruce. A positive effect occurs when the optimal dose of nitrogen is about 180 kg/ha. The positive effect of nitrogen on the photosynthesis of undergrowth *Picea abies* L is following:

1. The synthesis of photosynthetic pigments in needles and the intensity of photosynthesis increases.
2. The outflow of assimilates from the needles is accelerated and they supply actively working meristems (young needles, cambium, roots).

Along with external factors, internal factors, such as stomatal regulation (Berninger et al., 1996; Bolondinskiy, 2008) and metabolic processes (Troeng and Linder, 1982; Molchanov, 1986) also influencing photosynthesis.

Titova (2010, 2013) conducted research on the photosynthetic activity of needles of introduced fir trees in the arboretum Mountain-Taiga station of DVO RAS. The following species were studied: *Picea abies* (L.) *Karst., Picea pungens* Engelm., *Picea engelmanni* (Parry) Engelm., *Picea asperata* Mast., *Picea meyeri* Rehd. ex Wils. and the local species *Picea koraiensis* Nakai. Seasonal accumulation of plastid pigments, chlorophylls and carotenoids ratio in the needles of the introduced species were investigated. The similarity of quantitative characteristics of the pigment complex of introducents (*Picea pungens and Picea meyeri*) and far Eastern species (*Picea koraiensis*) was revealed. This indicates the balance of adaptation of the photosynthetic apparatus the Far Eastern species to light regime, climatic and natural conditions of the arboretum Mountain-Taiga station (Titova, 2013).

Seasonal growth dynamics and the content of photosynthetic pigments in the needles of various tree species deserve special attention (Titova, 2010; Kozina et al., 2015).

Pests and diseases cause poor plant growth, necrotic spots, and loss of foliage – all this reduces the synthesis of carbohydrates and the dimensions of the leaf blades are reduced. As a result, the photosynthetic activity of plants is suppressed (Kramer and Kozlovsky, 1983; Goryshina, 1989).

The influence of felling on the photosynthesis of woody plants is of particular interest to scientists. Konovalov and Zarubina (2011) studied the relationship between the intensity of fellings in birch forests bilberry on the metabolism of spruce undergrowth. Throughout the growing season, samples of spruce needles were taken to determine the intensity of potential photosynthesis. The intensity of potential photosynthesis was determined by the radiometric method with the concentration of radiocarbon in the working mixture 1%. Felling increased the influx of solar radiation under the forest canopy, activity of the root system increases in spruce undergrowth, the intensity of photosynthesis increases (Konovalov and Zarubina 2011).

Studies of variable and delayed fluorescence of chlorophyll are conducted more every year. The scientist interest in photoluminescent indicators defined by the possibility of creating express and non-contact methods, and in the future, remote methods of diagnosing the state of plants. Great prospects for the development of this area are determined by the integrality of photoluminescence indices (Stirbert et al., 2014; Ptushenko et al., 2014; Alieva et al., 2015).

Alieva and colleagues (2015) studied pigment indicators and fluorescent characteristics (fluorescence quantum yield, maximum fluorescence, quantum yield of photosynthesis) of lighting and shady leaves of *Acer platanoides* L. and *Fraxinus excelsior* depending on the degree of anthropogenic loading. There was a noted decrease in photosynthetic activity of the leaves, detected a reduction of maximum fluorescence in experienced samples.

Ptushenko and colleagues (2014) analyzed the variability of several parameters of chlorophyll fluorescence induction: Fmax = Fv | Fm (relative value of variable fluorescence, measured after a sufficiently long adaptation of the leaf to darkness); Gnpq (coefficient of non-photochemical quenching); Rfd ("viability factor") of the autumn leaves of ten tree species in the central part of Russia. The correlation of fluorescent indices of photosynthetic activity and chlorophyll content in leaves is shown on examples of *Tilia cordata* and *Sorbus aucuparia*. An adequate characteristic of the photochemical activity of leaves during autumn leaf color can be Fmax, while in summer, characteristics such as Gnpq or Rfd are more informative. A correlation between the value of Fmax, which characterizes the maximum photochemical activity of the photosystem II, and the "chromaticity coordinates" of a leaf, characterizing its color properties was revealed. The chromaticity coordinates were defined by the reflection spectra of light. They are quantitative characteristics of the color shades of the leaves. On the basis of the results, it is possible to create a visual expert assessment of the physiological state of the leaves (Ptushenko et al., 2014).

Thus, in recent decades, the photosynthesis was studied in hundreds of laboratories around the world. The information obtained created the basis for the prediction of photosynthesis and, accordingly, the productivity of plants. Approaches to mathematical modeling of the primary processes of photosynthesis are given in the next section of this chapter.

24.2.2 RESPIRATION

Plant respiration is an intracellular enzymatic multi-step process of oxidation of organic substances (mainly carbohydrates), formed during photosynthesis. This process is accompanied by the formation of various highly reactive metabolites for the synthesis of biomass and the release of energy, which is used for growth, development and other life processes of plants. The released energy is stored in the form of high-energy chemical compounds, mainly in the form of adenosine triphosphate (ATP) (Forest Encyclopedia, 1985). The respiratory activity of plants varies depending on the species of woody plants (Golovko et al., 2009). In woody plants, in addition to studying the respiration of leaves, a great interest is the respiration of non-assimilating organs: the stems and branches, since its value, is closely related to the outflow of assimilates from the crown (Cel'niker et al., 1993).

Two components can be distinguished in the respiration of plants: one is related to the formation of biomass, the second is to maintain the functioning of the cell. In young organs and tissues, where the share of growing cells is high, the bulk of the energy is used for growth (Golovko, 1985).

The respiration of the stems and branches during the growth period is characterized by a high intensity, which, per unit of surface, is much greater than the intensity of the respiration of the leaves (Cel'niker et al., 1993). The stem respiration has a daily rhythm and fluctuates during the vegetative period (Kunstle and Mitcherlich, 1976). The respiration depends on external environmental factors (mostly temperature), and is closely related to the growth of the stem in thickness (Negisi, 1981; Zabuga and Zabuga, 2005; Tatarinov et al., 2011; Bolondinskiy and Vilikainen, 2015).

Coniferous tree species are one of the most common forest-forming tree species in Russia. Scientists investigate not only the gas exchange of leaves, but also the gas exchange of trunks and branches of coniferous tree species (Zabuga and Zabuga, 2005; Masyagina et al., 2007; Tatarinov et al., 2011). The analysis carried out on groups of trees with a similar intensity of stem respiration showed that the average the stem respiration in a group of healthy trees was 1.5–1.6 times higher than the respiration of weakened and severely weakened trees (Masyagina et al., 2007).

The temporal dynamics of respiratory gas exchange CO_2 and skeletal branch radial growth in different parts of the crown of *Pinus sylvestris* was investigated in the forest-steppe districts of the Baikal region (Zabuga and Zabuga, 2005). The rate of CO_2-exchange was studied by means of the CO_2 analyzer, and the radial growth by microscopy. The annual and the average annual tree ring width (TRW) and the rate of CO_2 evolution of different-aged shoots decreased from the top of skeletal branches to their base, and those of even-aged shoots from upper to lower part of the crown. The radial increment of current year shoots at the branch top was the highest as compared with the other-aged shoots of fine branches. The average annual TRW of shoots correlated closely with the age ($r = -0.88$, $r = -0.92$, $p < 0.05$). Estimations of CO_2 evolution by the crown branches were made on the basis of the dependence of the axis shoots respiration on their radial growth. The calculated annual crown respiration was 2.9–5.7 kg CO_2, with current year, shoots respiration being 17.0–25.6 % of that of the crown. Total CO_2 evolution of the crown branches of *Pinus sylvestris* was 32.4–49.4% of the CO_2 amount evolved by the stem.

Bolondinsky and Vilikainen (2015) studied the twigs and stem respiration in two forms of *Betula pendula Roth* (*B. pendula var. pendula* and *B. pendula var. carelica*) during the vegetation. The research was carried out on 4 to 6-year old trees and 2 to 3-year old pot-grown seedlings. The main part of the measurements the intensity of dark respiration of twigs and stems was conducted using gasometrical system "Li–6200" (Li-Cor, USA) with the chamber original design, made of duralumin. The part of a twig or stem with a diameter more than 14 mm was placed into the chamber. The space between the sample and the chamber hole was sealed with sealant. The sensors of the assimilation chamber "Li–6200" measured the phased array, temperature and relative air humidity. As a result of the research, Bolondinskiy and Vilikainen (2015) revealed that before cambial growth (in June) mean respiration rates of twigs and stems in seedlings with diameters of 10–14 mm in Silver birch and Karelian birch were 5.3 and 4.6 $\mu mol\text{-}m^{-2}\text{-}s^{-1}$, respectively, during cambial growth (late June first half of August) 12.3 and 17.6 $\mu mol\text{-}m^{-2}\text{-}s^{-1}$, and during the period of growth inhibition (second half of August early September) 2.6 and 3.1 $\mu mol\text{-}m^{-2}\text{-}s^{-1}$. Maximal respiration rates (20 $\mu mol\text{-}m^{-2}\text{-}s^{-1}$ and 28 $\mu mol\text{-}m^{-2}\text{-}s^{-1}$ for Silver birch and Karelian birch, respectively) were registered in late July and were higher (in absolute value) than net CO_2 gas exchange of leaves. In about 30% of the measurements the respiration rate in Karelian birch was double that of Silver birch. Higher respiration rate in Karelian birch in comparison with Silver birch is mostly related to higher metabolic activity in the cambial zone and higher amount of living respiring tissues per unit of stem length in stems of the same width in the two birch forms.

Malishev and Golovko (2011) studied the temperature dependence of the respiration and metabolic heat production in the opening buds of *Syringa josikaea* and *S. vulgaris*. The metabolic heat production and respiration were measured isothermal microcalorimeters "Biotest-2." Respiration was measured as the heat flux from the test object and the reaction of the CO_2 released by the object with 0.4 M NaOH (Hansen et al., 1994). The heat dissipation rate increased with rise of temperature from 5 to 35°C, and the respiration rate was suppressed by high temperature. Buds of *S. vulgaris* grew more quickly than the buds of *S. josikaea*. The differences in the growth rate between these species were more pronounced in the temperature range of 20–30°C.

24.2.3   TRANSPIRATION

Transpiration is an important link in the chain of processes, which plays a significant role in the formation of forest stands. Transpiration of leaves is the most important factor in the water regime of plants because evapo-ration of water creates an energy gradient, which is the cause of the move-ment of water and lift xylem sap. This process also provides the stability of the internal temperature of the leaf and it prevents excessive pressure (Pakhomova and Bezuglov, 1980).

Water evaporates mainly from the leaf surface through stomata, but water loss can occur through the cuticle. This is true, especially for young leaves. Transpiration consists of two main processes: (a) water movement from leaf veins into the surface layers of the walls of the mesophyll; and

(b) evaporation of water from cell walls into intercellular spaces, with subsequent diffusion into the surrounding atmosphere.

Transpiration has daily and seasonal dynamics. Several approaches exist for the interpretation of the seasonal dynamics. Most of the researchers have linked the water loss with the combination of weather factors. However, a strong correlation is revealed only in woody plants with a sufficient level of moisture. The intensity of transpiration increases from spring to mid-summer and then it falls to the fall (Bihele et al., 1980; Akhmatov, 2016). It was found that good soil moisture increases the intensity of transpiration (Akhmatov, 2016).

The intensity of transpiration depends on a complex of external and internal factors. Temperature, humidity, light, wind, temperature and water regime of soil, and availability of moisture for plants are the most important external factors (Karasev, 2001).

Indicators of the water regime and the intensity of transpiration in coniferous forests were studied for *Pinus sylvestris* and *Picea obovata* in the Republic of Komi (Senkina, 2002, 2009), for *Picea abies* Karst. in the Arkhangelsk region (Ovsyannikova et al., 2012), for introduced pine species in the Republic of Mari El (Volzhanina, 2004) and South Primorye (Chernyshev, 2016).

Samples of needles were selected from the side branches of each studied trees. The branches were cut with pruning shears so that the sample remained current year needles and one-year needles (Ovsyannikova et al., 2012). Water deficit in the needles was calculated by the method of Stocker (Lear et al., 1974). The intensity of transpiration was determined by the method of fast weighing (Ivanov et al., 1950).

The intensity of transpiration of *Pinus sylvestris* is changed from 35 to 395 mg/g of wet weight per hour, of *Picea obovata* is from 20 to 280 mg/g of wet weight per hour in coniferous boreal forests of the Komi Republic. The most frequent intensity values of transpiration of *Pinus sylvestris* are in the range 200–300 mg/g of wet weight per hour, of *Picea obovata* 100–200 mg/g of wet weight per hour. The magnitude of the range of changes of the transpiration indicates the ability of plants to regulate water (Senkina, 2002).

The changes associated with phenological *phases* of plants have a significant impact on the intensity of the transpiration of coniferous woody plants. During the season, the differences in the intensity of transpiration was less for *Picea obovata* than *Pinus sylvestris*, and it made up about 10

%. The trend of reducing the intensity of transpiration with increasing age of the needles was observed. This is expressed especially clearly for *Picea obovata* (Senkina, 2002).

The tendency to decrease the intensity of transpiration with increasing age of needles is observed. This is especially pronounced for spruce (Senkina, 2002). Akhmatov and colleagues (1999) introduced changes in the methodology for studying transpiration. They studied the seasonal dynamics of transpiration intensity for 15 tree species and 16 species of shrubs using this methodology (Akhmatov, 2016).

Mushinskaya and colleagues (2007) conducted a study of transpiration for three species of the genus *Populus* L.: *P. pyramidalis* Roz., *P. balsamifera* L., *P. nigra* L., which grow in Orenburg (Russia). Minimum values of transpiration were recorded in the morning, and highest – in the afternoon and evening. The most intensive transpiration is characterized by *P. nigra*, and the lowest intensity of transpiration was detected in *P. balsamifera*.

The role of transpiration in the formation of the water regime in oak forests of Primorye is covered in the works of Siritsa (2009). The peculiarities of the water regime of *Betula pendula* Roth during the vegetation period have been investigated by Amosova and Feklistov (2010) and by Rybakova and Kulagin (2015). The intensity of transpiration of *Betula pendula* Roth depends on the age and condition of the tree, varies in different parts of the crown and during the summer (Amosova and Feklistov, 2010), which is also typical for other woody plants. The influence of pollution on transpiration intensity, water deficit and relative water content in the assimilation organs of *Betula pendula* was established (Rybakova and Kulagin, 2015). The increase in pollution has a greater effect on the intensity of transpiration and water deficiency and, it has a lesser extent, on the total water content in leaves.

24.3 STUDIES OF THE WOODY PLANT'S ADAPTATION TO EXTREME ENVIRONMENTAL CONDITIONS AND POLLUTION

The ability of plants to adapt to environmental conditions is an important factor, which determines their resistance. Studies on the adaptation of plant organisms to environmental change are relevant in connection with the deterioration of the ecological state of the biosphere. It is understood that the resistance of plants to adverse abiotic and biotic factors is the result of the operation of a lot of diverse mechanisms operating at different levels of biological organization (Titova et al., 2011; Kaznina et al., 2016).

There is the notion that the ratio of N:P:K is a homeostatic indicator of the functional state of the plant organism. This question is explored for a variety of woody plants. Special researches in order to assess the intensity of absorption of mineral substances (adaptation to extreme environmental conditions) are carried out for birch (Gorbunova, 2012) and pine (Habarova et al., 2015).

In technogenic conditions, the homeostatis of the elemental composition of plants is disturbed (Sukhareva and Lukina, 2004; Buharina et al., 2007; Sukhareva, 2012). Nitrogen metabolism of plants is the most changed (Buharina et al., 2007). The distribution of the main elements in the structural parts of the plants and the autumn physiological outflow of elements from leaves to the resting shoots change (Khrustaleva, 2002; Buharina et al., 2007).

Many researchers have observed an increase of the NPK concentration in the leaves and needles of trees under the influence of pollution (Lukina and Nikonov, 1991; Neverova and Kolmogorova, 2003). It was assumed that the increase of the concentration of these elements in conifers due to the constant shedding of needles (not only in phenological terms) and the outflow of mobile elements from falling needles to ones still on the tree (Lukina and Nikonov, 1991). However, it is established that an increased concentration of nitrogen in leaves associated with the ability of plants in stress conditions to increase the content of free amino acids and ability to be involved in the metabolism of nitrogen-containing gaseous pollutants from the air, such as nitrogen oxides and ammonia (Vasfilov, 2003). The excess nitrogen is toxic to plants, as it causes inhibition of photosynthesis through suppression of the synthesis of chlorophyll (Buharina et al., 2007).

Sukhareva (2012, 2013) investigated the effect of emissions from copper-nickel production on the chemical composition of assimilating organs of pine and birch. It was revealed that birch leaves contain higher concentrations of Cu, Ni and S than pine needles. An insufficient provision of pine needles with phosphorus and potassium was found (Sukhareva, 2012, 2013).

Researchers pay special attention to the peculiarities of the water regime of woody plants in the urban environment (Rybakova and Kulagin, 2015) and industrial pollution (Sazonova et al., 2009). *Betula pendula*

Roth, *Pinus sylvestris* L. and *Picea obovata* Ledeb. are the most studied woody plants (Sazonova et al., 2009; Rybakova and Kulagin, 2015). It was found that the increase in pollution has a greater impact on the intensity of transpiration and water deficits and to a lesser extent on the total water content in the leaves (Rybakova and Kulagin, 2015).

Photosynthesis and respiration are sensitive to various stresses (Rakhmankulova, 2002). The intensity of photosynthesis is reduced in various woody plants (pine, spruce, linden, birch, lilac, ash) in urban areas with a high degree of pollution (Neverova and Candekova, 2010). The aerial technogenic emissions influence on the photosynthetic apparatus was considered for *Pinus sylvestris* L (Bazhenov and Shavnin, 1994), *Picea obovata* Ledeb. and *P. abies* (L.) karst (Tarkhanov and Biryukov, 2014). The content of photosynthetic pigments, especially chlorophyll, increases in annual spruce needles for spruce stands of Northern taiga under atmospheric pollution. It is revealed that the intensity of photosynthesis, content of chlorophyll and carotenoids in annual needles of Scots pine significantly reduced in pine forests with bilberry and sphagnum when approaching the source of emissions (Tarkhanov and Biryukov, 2014).

Adaptive properties of woody plants enable to evaluate the air pollution of major cities. For example, morphometric and chemical characteristics of needles of *Picea obovata* Ledeb. are used in the Krasnoyarsk city (Esyakova and Stepen, 2008, 2015).

The problem specific features of organs of woody plants (leaves and bark) to accumulate cadmium entering through atmospheric precipitation are considered for Orenburg city. It was found that various tree species statistically differ in their ability to accumulate cadmium. Trees can be placed in a row according to their ability to accumulate cadmium: *Betula pendula* L. > *Populus nigra* L. > *Ulmus pumula* L. > *Fraxinus excelsior* L. (Naumenko et al., 2015).

We can conclude that the study of the adaptation of woody plants to urban conditions and industrial pollution is urgent. Research covers a wide range of problem areas. Studies of plant adaptation to extreme environmental conditions are less common and many unresolved issues remained.

24.4 MODELING OF THE WOODY PLANT PRODUCTION

At present, it is generally accepted that modeling is an excellent methodological approach for the study of woody plants and forest communities (Guts and

Volodchenkova, 2012). The models reflect one of the views on the problem and consciously simplify reality (Risch et al., 2005).

Photosynthesis is a popular object of mathematical modeling. A complex of mathematical models was proposed for testing null hypotheses about the organization and regulation of photosynthesis and the assessment of active factors for electron transfer during photosynthesis. Systems of ordinary differential equations are used most often. Modern models usually consist of a few dozen differential equations (Blankenship, 2014; Rubin and Riznichenko, 2014). However, such models have limitations (Maslakov et al., 2016).

Monte Carlo method is becoming increasingly popular for modeling processes in the living cell. It is associated with the development of computing possibilities. Monte Carlo method involves numerical methods which are based on acquiring a large number of realizations of a random process. Modeling using the Monte Carlo method involves a simulation of the behavior of elementary parts of a complex system (Maslakov et al., 2016).

It is generally accepted that biological productivity can act as the main characteristic of plant communities (Usoltsev et al., 2014). Therefore, the importance of studying biological productivity is difficult to overestimate, especially in the context of global climate change (Jiang et al., 1999).

Forest growth models are made for individual trees as well as the whole of a forest stand. The book by Vanclay (1994) is an introduction to the growth modeling. Examples of forest growth models can be found at Munro (1974). Process models examine growth processes depending on conditions (Hope, 2003). For example, Laubhann et al., (2009) modeled the growth of forest depending on temperature and nitrogen.

Eco-physiological simulation models which are built on the concept of distribution of the products of photosynthesis (assimilates) between the parts of the tree, are fairly widespread. The concept of the functional equilibrium found the widest application (Davidson, 1969). A modification of the model of functional equilibrium is built on the ranking of individual parts of the tree by priority of receiving assimilates. The weight of the trunk is a measure of the excess of assimilates which remain after satisfaction of the costs necessary for the growth of roots, branches, pine needles, fruits to self-defense and respiration (Waring, 1980). Such models are developed at the level of the tree, and they are extrapolated to forest ecosystems to estimate their production capacity (Rachko, 1978; Palumets, 1990).

In recent years, in connection with the increasing awareness of the biosphere role of forest cover in the construction of functional models, there will the transition from the level of individual plantations on landscape level (Kimmins, 1986). These models include a description of the fundamental processes of energy, carbon, nitrogen and water exchanges in vegetation and ecosystem response to climate change and pollution. The basis of such models is the unification of fundamental plant biology with ecosystem function to simulate ecosystem processes, including tree canopy photosynthesis, transpiration, changes in soil moisture, the dynamics of carbon and nutrients. The model focuses on using remote sensing and GIS technologies to combine data on the structure of vegetation with the climate data and characteristics of the habitats. FOREST-BGC (Running and Coughlan, 1988; Running, 1994), CENTURY (Parton et al., 1992), TEM (Raich et al., 1991), BIOME 1 and BIOME 3 (Prentice et al., 1992; Haxeltine and Prentice, 1995), DOLY (Woodward et al., 1995), MAPSS (Neilson and Marks, 1994) are the models in this category.

The General approach of determination of productivity (NPP) is based on using land cover maps in conjunction with patterns in vegetation productivity on the basis of GIS-technologies (Jiang et al., 1999). LAI-index (the ratio of leaf area to the area occupied by plantation) is a highly informative characterization of forest canopy associated with its energy and mass transfer, and is evaluated with satellite sensors with high resolution in wide areas (Running et al., 1986).

This index is introduced in the model as the main independent variable for calculating the processes of light interception of the canopy, transpiration, photosynthesis, growth and carbon sequestration in above-ground and underground area. However, the sheer volume and complexity of incoming data make it difficult to use traditional mathematical methods. Lankin and Ivanova (2015) propose to use dynamic neural networks, which are extremely flexible and efficient tool for data analysis. High flexibility, precision and high simulation efficiency is important features of neural networks which allows solving wide range of tasks using the same mathematical algorithms (Lankin et al., 2012). Modeling and forecasting using artificial neural networks is based on samples of the source data required for training neural networks. The authors (Lankin et al., 2012) obtained good agreement between theoretical and experimental data.

For Komarov and his colleagues (2003), the challenge was to propose a system of models for the study of forest ecosystems. This model is called

EFIMOD. This was achieved by applying a species-specific parameter: the maximum net biological productivity per unit mass of photosynthesizing organs. This was achieved by applying a species-specific parameter: the maximum net biological productivity per unit mass of photosynthesizing organs. The proposed model takes into account such interactions between plants as: competition for light and competition for available nitrogen in the soil. It allowed us to predict the direction of reforestation.

EFIMOD 2 is designed for modeling (including long-term) of the dynamics of natural and controlled tree stand in different climatic conditions and silvicultural regimes. A feature of the model is the ability to determine the dendrometric characteristics for an individual woody plant (Komarov et al., 2003).

KEYWORDS

- **dehydrator**
- **inductively coupled plasma mass spectrometry**
- **photosynthesis**
- **physiological processes**
- **respiration**
- **x-ray fluorescence**

REFERENCES

Adamenko, V. N., Zhuravleva, E. L., & Chetverikov, A. F., (1982). Chemical composition of annual rings of trees and the state of the environment. *Reports of the Academy of Sciences of the USSR, 265*(2), 507–512.

Akhmatov, M. K., (2016). Seasonal dynamics of transpiration intensity of woody plants introduced in the Chui Valley. *Science of the XXI Century: Problems and Prospects, 1*, 8–14.

Akhmatov, M. K., Oskonbaeva, R. K., & Shpota, L. A., (1999). A new technique for the use of field transpirometers in determining the intensity of transpiration in shrubs and lianas. *Introduction and Acclimatization of Plants in Kyrgyzstan*, pp. 120–123.

Alieva, M. Yu, Mammaev A. T., & Magomedova, M. H.-M., (2015). The photosynthetic characteristics of lighting and shady leaves of woody plants in Makhachkala city.

Proceedings of the Samara Scientific Center of the Russian Academy of Sciences, 17, 67–71.

Amosova, I. B., & Feklistov, P. A., (2010). Water mode assimilation of the device silver birch (*Betula pendula* Roth). *Forestry Bulletin, 6,* 26–29.

Berninger, F., Makela, A., & Hari, P., (1996). Optimal control of gas exchange during drought: Empirical evidence. *Ann. Bot., 77,* 469–476.

Bihele, Z. N., Moldau, H. A., & Ross, Y. K., (1980). *Mathematical Modeling of Transpiration and Photosynthesis in the Absence of Soil Moisture* (p. 224). Leningrad: Gidrometeoizdat.

Blankenship, R. E., (2014). *Molecular Mechanisms of Photosynthesis* (2nd edn., p. 322). Wiley-Blackwell.

Bobkova, K. S., (1987). *Biological Efficiency of Coniferous Forests of the European Northeast* (p. 156). Leningrad, Nauka Publ.

Bolondinskiy, V. K., (2008). *Ustichnaya Regulation of Photosynthesis in Scots Pine* (pp. 15–17). Fundamental and applied problems of botany in the beginning of the XXI century: Materials conf. Petrozavodsk, KarRC RAS.

Bolondinskiy, V. K., (2010). Research of dependence of photosynthesis on the intensity of solar radiation, air temperature, and humidity in Karelian (curly) and silver birch plants. *Transactions of Karelian Research Centre of the Russian Academy of Sciences, 2,* 3–9.

Bolondinskiy, V. K., & Vilikainen, L. M., (1987). Photosynthesis of Scots pine in various types of forest. *Ecophysiological Studies of Woody Plants* (pp. 77–85). Petrozavodsk, Karelian branch of the Academy of Sciences of the USSR.

Bazhenov, A. V., & Shavnin, S. A., (1994). Estimation of the degree of photosynthesis damage of Scots pine by aerotechnogenic emissions. *Russian Journal of Ecology, 4,* 89–91.

Bolondinskiy, V. K., & Vilikainen, L. M., (2015). *Research of Respiration of Twigs and Stems in Karelian Birch and Silver Birch* (Vol. 12, pp. 66–79). Transactions of the Karelian Branch of the USSR Academy of Science. doi: 10.17076/eb249.

Buharina, I. L., & Dvoeglazova, T. M., (2010). *Bioecological Features of Herbaceous and Woody Plants in Urban Plantations* (p. 184). Izhevsk, Udmurt University Publishing House.

Buharina, I. L., Dvoeglazova, T. M., & Vedernikov, K. E., (2007). *Ecological and Biological Features of Woody Plants in an Urbanized Environment* (p. 216). Izhevsk, Izhevsk State Agricultural Academy.

Cel'niker, Y. L., Malkina, I. S., Kovalev, A. G., Chmora, S. N., Mamaev, V. V., & Molchanov, A. G., (1993). *Growth and CO$_2$ Exchange in Woody Plants* (p. 256). Moscow, Nakua Publ.

Chernyshev, M. V., (2016). Comparative anatomical-physiological indices for adaptation estimation of four coniferous species to conditions of South Primorye. *International Research Journal, 5*(47), 100–103.

Davidson, R. L., (1969). Effect of root/leaf temperature differentials on root/shoot ratios in some pasture grasses and clover. *Ann. Bot., 33*(131), 561–569.

Demakov, Y. P., Isaev, A. V., & Shvecov, A. M., (2013). Consumption and removal of ash constituents from wooden plants in the inundated biotope. *Vestnik of Volga State University of Technology. Series: Forest, Ecology, Nature Management, 1,* 36–49.

Demakov, Y. P., & Shvecov, S. M., (2013) The content of ash elements in annual layers of pine trees in lakeside biotopes of the national park "Mari Chodra". *Eco-Potential, 3-4,* 128–136.

Esyakova, O. A., & Stepen, R. A., (2008). Indication of atmospheric pollution in Krasnoyarsk on the morphometric and chemical parameters of Siberian pine needles. *Khimiia Rastitel'nogo syr'ia (Chemistry of Plant Raw Material), 1,* 143–148.

Esyakova, O. A., & Stepen, R. A., (2015). Volumetric development method of measuring the needles in bioindication air pollution. *Forestry Bulletin, 2,* 11–14.

Forest Encyclopedia, (1985). (p. 563). Moscow, Sov. Encyclopedia.

Golovko, T. K. (1985). The system of indicators in the study of the role of respiration in the production process of plants. *Russian Journal of Plant Physiology, 32*(5), 1004–1013.

Golovko, T. K., Dalke, I. V., Tabalenkova, G. N., & Garmash, E. V., (2009). Respiration of plants of the Subpolar Urals. *Botanicheskij Zhurnal (Bot. J.), 94*(8), 1216–1226.

Gorbunova, V. D., (2012). Analysis of macroelements content in the leaves of white birches and soils along the high gradient in the South Urals. Vestnik of the Orenburg State University. *6*(38), 193–196.

Goryshina, T. K., (1989). *Photosynthetic Apparatus of Plants and Environmental Conditions* (p. 202). Leningrad, Leningrad State University.

Grachev, A. M., Vaganov, E. A., Leavitt, S. W., Panyushkina, I. P., Chebykin, E. P., Shishov, V. V., et al., (2013). Methodology for development of a 600-year tree-ring multi-element record for larch from the Taymir peninsula, Russia. *J. Sib. Federal Univ. Biology, 6*(1), 61–72.

Guts, A. K., & Volodchenkova, L. A., (2012). *Cybernetics of Forest Ecosystem Catastrophes* (p. 220). Omsk, KAN Publ.

Habarova, E. P., Feklistov, P. A Kosheleva, A. E., (2015). Contents of mineral elements in the dying off needles of Scotch pine on drained areas. *Forestry Bulletin, 2*(19), 15–20.

Hansen, L. D., Hopkin, M. S., Rank, D. R., Anekonda, T. S., Breidenbuch, R. W., & Criddle, R. S., (1994). The relation between plant growth and respiration: A thermodynamic model. *Planta, 194*(1), 77–85.

Haxeltine, A., & Prentice, I. C., (1995). BIOME-3: An equilibrium terrestrial biosphere model based on ecophysiological constraints, resource availability, and competition among plant functional types. *Global Biogeochem. Cycles, 10,* 693–709.

Hope, J. C. E., (2003). Modeling forest landscape dynamics in Glen Affric, northern Scotland. *PhD Thesis* (p. 317). University of Stirling.

Ilkun, G. M., (1978). *Pollutants of the Atmosphere and Plants* (p. 246). Kiev, Naukova Dumka.

Ivanov, L. A., Silina, A. A., & Cel'niker, Yu. L., (1950). About a rapid weighing method for determining transpiration *in vivo. Botanicheskii Zhurnal, 35*(2), 171–185.

Jiang, H., Apps, M. J., Zhang, Y., Peng, C., & Woodard, P. M., (1999). Modeling the spatial pattern of net primary productivity in Chinese forests. *Ecological Modeling, 122,* 275–288.

Karasev, V. N., (2001). *Plant Physiology* (p. 263). Yoshkar-Ola, Mari State Technical University.

Katrushenko, I. V., (1983). *Daily Dynamics of CO$_2$-Gas Exchange in the Canopy Crowns of Spruce Stands: Factors of Regulation of Spruce Ecosystems* (pp. 112–118). Leningrad, Nauka Publ.

Kazimirov, N. I., & Morozova, R. M., (1973). *Biological Circulation of Substances Infer Groves of Karelia* (p. 175). Leningrad, Nauka Publ.

Kaznina, N. M., Batova, Y. V., Titov, A. F., & Laidinen, G. F., (2016). Role of antioxidant system components in adaptation of *Elytrigia repens* (L.) 'Nevski' to cadmium. *Transactions of Karelian Research Centre of Russian Academy of Science. Experimental Biology*, *11*, 17–26.

Kholoptseva, E. S., & Chernobrovkina, N. P., (2009). Effect of nitrogen, boron and *Lupinus angustifolius L* on the growth and mineral nutrition of Scots pine. *Russian Forest Sciences*, *1*, 25–32.

Kholoptseva, E. S., Drozdov, S. N., Sazonova, T. A., & Khilkov, N. I., (2012). Influence of soil temperature and other environmental factors on photosynthesis of birch seedlings on the surface. *Proceedings of Petrozavodsk State University*, *8*, 28–31.

Khrustaleva, M.A., (2002). Ecogeochemistry of morainic landscapes of the center of Russian Plains (p. 314) Moskow, Tehpoligratsentr.

Kimmins, J. P., (1986). *Forest Ecology* (p. 930). New York: Macmillan Publishing Company.

Komarov, A., Chertov, O., Zudin, S., Nadporozhskaya, M., Mikhailov, A., Bykhovets, S., Zudina, E., & Zoubkova, E., (2003). EFIMOD 2–a model of growth and cycling of elements in boreal forest ecosystems. *Ecological Modeling*, *170*(2/3), 373–392. doi: 10.1016/s0304-3800(03)00240-0

Konovalov, V. N., & Zarubina, L. V., (2011). Influence of forest felling on photosynthesis and ^{14}C-otosynthates discharge in spruce understory of blueberry birch forests. *Arctic Environmental Research*, *3*, 49–54.

Kozina, L. V., Titova, M. S., Ivaschenko, E. A., Rezinkina, G. A., Karasev, V. E., & Mirochnik, A. G., (2015). Growth and photosynthetic process of seedlings coniferous species under the fluorescent envelope. *Modern Problems of Science and Education, 3*, URL: www.science-education.ru/123-19844 (accessed on 14 January 2020).

Kramer, P. D., & Kozlovsky, T. T., (1983). *Physiology of Woody Plants* (p. 462, 464). Moscow: Publishing House of Forest Industry.

Kunstle, E., & Mitcherlich, G. (1976). Photosynthesis, respiration and perspiration in a mixed population in the Black Forest. 3. Atmug. *General Forest and Jagdztg, 147*(9), 169–177.

Kurets, V. K., & Drozdov, S. N., (2008). Laboratory evaluation of photosynthetic activity of seedlings of woody plants. *Russian Forest Sciences, 3*, 57–78.

Lankin, Y. P., & Ivanova, N. S., (2015). Methodological problems in the modeling of ecosystems and ways of solutions. *International Journal of Bio-Resource and Stress Management, 6*(5), 631–638. doi: 10.5958/0976-4038.2015.00098.6.

Lankin, Y. P., Mokogon, D. A., & Tereshin, S. V., (2012). Adaptive modeling of planetary processes on the basis of satellite data. *Modern Problems of Science and Education, 6*. URL: https://science-education.ru/en/article/view?id = 7136 (accessed on 14 January 2020).

Laubhann, D., Sterba, H., Reinds, G., & De Vries, W., (2009). The impact of atmospheric deposition and climate on forest growth in European monitoring plots: An individual tree growth model. *Forest Ecology and Management, 258*, 1751–1761.

Lear, C., Polster, G., & Fiedler, G., (1974). *Physiology of Woody Plants* (p. 421). Moskow.

Lebedev, E. V., (2015). Photosynthesis, mineral nutrition and biological productivity pine stands of different yield class in Belarus in ontogenesis. *RUDN Journal of Ecology and Life Safety*, *4*, 37–45.

Lugansky, N. A., Zalesov, S. V., & Lugansky, V. N., (2010). *Lesovedenie* (p. 432). Ekaterinburg, Publishing House. Ural State Forestry University.

Lukina, I. V., & Nikonov, V. V., (1991). Determination of primary productivity and condition of technogenically damaged stands. *Protection and Rational Use of Forest Resources* (pp. 42–44). Moskow.

Malishev, R. V., & Golovko, T. K., (2011). The respiration and energy balance in woody plant buds after breaking. *Bulletin of the Institute of Biology of the Komi Scientific Center of the Ural Branch of the RAS*, *7/8*, 25–28.

Maslakov, A. S., Antal, T. K., Riznichenko, G. Y., & Rubin, A. B., (2016). Modeling of primary photosynthetic processes using a kinetic Monte Carlo method. *Biophysics*, *61*(3), 464–677.

Masyagina, O. V., Prokushkin, S. G., & Ivanova, G. A., (2007). Influence of fires on the intensity of respiration of the pine trunks (*Pinus silvestris L.*). *Conifers of the Boreal Zone*, *XXIV*(1), 82–91.

Molchanov, A. G., (1986). The ratio of photosynthesis and transpiration in Scots pine in southern taiga. *Russian Forest Sciences*, *4*, 76–82.

Molchanov, A. G., (2012). Intensity of photosynthesis in *Quercus robur* phenological forms under moisture deficit. *Russian Forest Sciences*, *4*, 31–38.

Molchanov, A. G., (2015). Variability of photosynthetic light curves of tree species. *Russian Forest Sciences*, *1*, 20–26.

Monsi, M., & Saeki, T. (1953). About the light factor in plant societies and its importance for die Stoff production. *Jap. J. Bot.*, *14*(1), 22–52.

Munro, D. D., (1974). Forest growth models: A prognosis. In: Fries, J., (ed.), *Growth Models for Tree and Stand Simulation* (pp. 7–21). Royal College of Forestry, Stockholm, Sweden.

Mushinskaya, O. A., Ryabinina, Z. N., & Mushinskaya, N. I., (2007). *Transpiration as an Integral Part of the Water Regime of Plants and its Study in Species of the Genus Populus L.* (Vol. 6, pp. 95–99). Vestnik of the Orenburg State University.

Naumenko, O. A., Sokolova, O. Y., & Bibartseva, E. V., (2015). *Characteristics of Specific features of the Cadmium-Binding Capacity of Woody Plants in the City of Orenburg* (Vol. 10, No. 185, pp. 225–228). Vestnik of the Orenburg State University.

Negisi, K., (1981). Diurnal and seasonal fluctuations in the stem bark respiration of standing *Quercus myrsinaefolia* tree. *J. Japan. Forest Soc.*, *63*(7), 235–241.

Neilson, R. P., & Marks, D., (1994). A global perspective of regional vegetation and hydrologic sensitivities from climatic change. *Journal of Vegetation Science*, *5*(5), 715–730.

Neverova, O. A., & Candekova, O. L., (2010). Photosynthetic ability of woody plants as an indicator of the cumulative air pollution of the urban environment. *Sibirskiy Ekologicheskiy Zhurnal*, *17*(2), 193–196.

Neverova, O. A., & Kolmogorova, E. U., (2003). Woody plants and urban environment: Ecological and biotechnological aspects. *Novosibirsk, Science*, p. 222.

Ovsyannikova, N. V., Feklistov, P. A., Volkova, N. V., Mochalov, B. A., Melekhov, V. I., & Drozdov, I. I., (2012). Indicators of the water regime of pine needles in the bilberry forest type. *Forestry Bulletin*, *3*, 24–29.

Pakhomova, G. I., & Bezuglov, V. K., (1980). *Water Regime of Plants* (p. 252). Kazan.

Palumets, Y. K., (1990). Experience in modeling the distribution of spruce biomass. *Lesovedenie, 3*, 43–48.

Parton, W. J., Mckeown, R., Kirchner, V., & Ojima, D. S., (1992). *Century User's Manual* (p. 289). Colorado State University, Natural Resource Ecology Laboratory: Ft. Collins.

Prentice, I. C., Cramer, W., Harrison, S. P., Leemans, R., Monserud, R. A., & Solomon, A. M., (1992). A global biome model based on plant physiology and dominance, soil properties and climate. *Journal of Biogeography, 19,* 117–134.

Rachko, P., (1978). Imitative model of tree growth: Construction of the model. *Russian Journal of Biology, 39*(4), 563–571.

Raich, J. W., Rastetter, E. B., Melillo, J. M., & Kicklighter, D. W., (1991). Potential net primary productivity in South America: Application of a global model. *Ecol. Appl., 1,* 399–429.

Rakhmankulova, Z. F., (2002). Interrelation of photosynthesis and respiration of an entire plant in normal and unfavorable external conditions. *Biology Bulletin Reviews, 63,* 3, 44–53.

Risch, A., Heiri, C., & Bugmann, H., (2005). Simulating structural forest patterns with a forest gap model: A model evaluation. *Ecological Modeling, 18,* 161–172.

Rubin, A. B., (2005). Biophysics of photosynthesis and methods of ecological monitoring. *Technologies of Living Systems, 1/2,* 425–453.

Rubin, A. B., & Riznichenko, G. Y., (2014). *Mathematical Biophysics.* Springer. doi: 10.1007/978-1-4614-8702-9.

Running, S. W., & Coughlan, J. C., (1998). A general model of forest ecosystem processes for regional applications. I. Hydrologic balance, canopy gas exchange and primary production processes. *Ecological Modeling, 42,* 125–154.

Running, S. W., (1994). Testing forest-BGC ecosystem process simulations across a climatic gradient in Oregon. *Ecol. Appl., 4,* 238–247.

Running, S. W., Peterson, D. L., Spanner, M. A., & Teuber, K. B., (1986). Remote sensing of coniferous forest leaf area. *Ecology, 67,* 273–276.

Rybakova, E. A., & Kulagin, A. A., (2015). Peculiarities of the water regime in birch (*Betula pendula* Roth) leaves during the vegetative period in the territory of Ufa city Bashkortostan republic. *Proceedings of the Samara Scientific Center of the Russian Academy of Sciences, 5,* 193–196.

Sazonova, T. A., Kolosova, S. V., & Isaeva, L. G., (2007). Water regime of *Pinus sylvestris and Picea obovata* (Pinaceae) under industrial pollution. *Botanicheskiy Zhurnal, 92*(5), 740–750.

Sazonova, T. A., Pridacha, V. B., & Kolosova, S. V., (2009). On the content of mineral elements in xylem sap. *Rastitel'nye Resursy, 45*(1), 113–121.

Senkina, S. N., (2002). *Moisture in the Plant Production Process* (Vol. 11, pp. 2–5). Bulletin of the Institute of Biology of the Komi Scientific Center of the Ural Branch of the RAS.

Senkina, S. N., (2009). Transpiration and stomatal resistance of Scotch pine in different growing conditions. *Lesnoy Zhurnal (Forestry Journal), 6,* 45–52.

Shiyatov, S. G., (1986). *Dendrochronology of the Upper Forest Boundary in the Urals* (p. 136). Moscow, Nauka Publ.

Siritsa, M. V., (2009). The role of transpiration in the assessment of water quality of oak forests. The bulletin of the Far Eastern Federal University. *Economics and Management, 1,* 70–75.

Sukhareva, T. A., (2012). Elemental composition of leaves of woody plants in conditions of technogenic pollution. *Chemistry for Sustainable Development, 20,* 369–376.

Suvorova, G. G., (2009). *Photosynthesis of Coniferous Tree in Siberia* (p. 195). Novosibirsk, Geo.

Suvorova, G. G., Oskorbina, M. V., Kopytova, L. D., Oskolkov, V. A., & Yankova, L. S., (2012). Unit and total photosynthetic productivity of stands of Irkutsk region. *Proceedings of the Samara Scientific Center of the Russian Academy of Sciences,* 1554–1557.

Tatarinov, F. A., Molchanov, A. G., & Kurbatova, Y. A., (2011). The influence of weather and soil conditions on stem respiration in spruce forest at south-west of Valdai hill. *Structural and functional deviations from normal growth and development of plants under the influence of environmental factors. Petrozavodsk,* pp. 346–351.

Titov, A. cF., Talanova, V. cV., & Kaznina, N. M., (2011). Physiological basis of plant resistance to heavy metals (p. 77). Petrozavodsk, Karelian Research Center of the RAS.

Titova, M. S., (2010). Seasonal dynamics of the pigments availability in needles of Siberian pine (*Pinus sibirica*) and Korean pine (*Pinus koraiensis*). *The Bulletin of KrasGAU, 8,* 77–81.

Titova, M. S., (2013). Features photosynthetic activity needles introduced species Picea A. Dietr., in the arboretum mountain taiga station. *Fundamental Research, 11,* 128–132.

Troeng, E., & Linder, S., (1982). Gas exchange in a 20-year-old stand of Scots Pine II. Variation in net photosynthesis and transpiration within and between trees. *Physiol. Plant, 54*(1), 15–23.

Ptushenko, V. V., Ptushenko, O. S., & Tikhonov, A.N., (2014). Fluorescence induction, chlorophyll content and color characteristics of leaves as indicators of aging of the photosynthetic apparatus of woody plants. *Biochemistry, 79*(3), 338–352.

Tyukavina, O. N., & Kunnikov, F. A., (2015). The contents of mineral elements in the phytomass of scots pine and balsam poplar wood in Arhangelsk. *Arctic Eviromental Research, 3,* 80–86.

Usoltsev, V. A., Chasovskikh, V. P., & Noricina, Y. V., (2014). Simulation modeling of forest ecosystems and the problem of replacing fossil fuels with green energy. *Eco-Potential, 4*(8), 16–40.

Vaganov, E. A., Grachev, A. M., Shishov, V. V., Menyailo, O. V., Knorre, A. A., Panyushkina, I. P., Leavitt, S. W., & Chebykin, E. P., (2013). Elemental composition of tree rings: A new perspective in biogeochemistry. *Doklady Biological Sciences, 453*(1), 375–379.

Vanclay, J., (1994). Modeling forest growth and yield: Applications to mixed tropical forests sensitivities from climatic change. *J. Veget. Sci., l*(5), 715–730.

Vasfilov, S. P., (2003). Possible ways of negative influence of acid gases on plants. *Biology Bulletin Reviews, 64*(2), 146–159.

Veretennikov, A. V., (1987). *Physiology of Plants with the Fundamentals of Biochemistry* (p. 256). Voronezh, VSU.

Vinokurova, R. I., & Lobanova, O. V., (2011). Specificity of distribution of macrocells in parts of wood plants of spruce-fir forests in the republic of Mari El. *Vestnik of Volga State University of Technology. Series: Forest Ecology: Nature Management, 2,* 76–83.

Volzhanina, E. M., (2004). *Estimation of the Resistance of Introduced Species of Pine Trees According to Indicators of Water Retention Capacity of Pine Needles* (Vol. 8, No. 2, pp. 94–99). Vestnik of Lobachevsky University of Nizhni Novgorod.

Waring, R. H., (1980). *Site, Leaf Area and Phytomass Production in Trees* (pp. 125–135). Mountain environments and subalpine tree growth. Techn. Paper No. 70. Forest. Res. Institute. N.Z. Forest Serv.

Woodward, F. I., Smith, T. M., & Emanuel, W. R., (1995). A global land primary productivity and phytogeography model. *Global Biogeochem. Cycles, 9,* 471–490.

Yalynskaya, E. E., & Chernobrovkina, N. P., (2008). Stability of pine seedlings to snow Shute, as an integral indicator of functional diagnostics of boron plants. *Forestry Bulletin, 6,* 160–163.

Zabuga, V. F., & Zabuga, G. A., (2005). The estimation of respiratory expenses of *Pinus sylvestris (Pinaceae)* branches by their radial growth. *Botanicheskii Zhurnal, 90*(12), 1867–1878.

Zarubina, L. V., (2016). The effect of nitrogen on photosynthesis and growth of spruce in the birch forests (the case of Lomovoye station). *Vestnik of Northern (Arctic) Federal University. Series: "Natural Sciences," 2,* 51–64.

Zavyalova, N. S., (1976). Light curves of photosynthesis of pine and spruce growth in the subzone of the southern taiga of the trans-Urals. *Ecological and Physiological Studies of Coniferous Tree Species in the Urals* (pp. 24–40). Sverdlovsk, UFAN USSR.

Index

Printed in the United States
by Baker & Taylor Publisher Services

Printed in the United States
by Baker & Taylor Publisher Services